A Scientific Autobiography:
S. CHANDRASEKHAR

A Scientific Autobiography:
S. CHANDRASEKHAR

Edited by

Kameshwar C. Wali

Syracuse University, USA

 **World Scientific**

NEW JERSEY · LONDON · SINGAPORE · BEIJING · SHANGHAI · HONG KONG · TAIPEI · CHENNAI

Published by

World Scientific Publishing Co. Pte. Ltd.

5 Toh Tuck Link, Singapore 596224

USA office: 27 Warren Street, Suite 401-402, Hackensack, NJ 07601

UK office: 57 Shelton Street, Covent Garden, London WC2H 9HE

British Library Cataloguing-in-Publication Data
A catalogue record for this book is available from the British Library.

Cover credit: Subrahmanyan Chandrasekhar in his office at the University of Chicago, standing beside a picture of Sir Isaac Newton placed on the volumes of the *Principia Mathematica*, March 1991. Photograph courtesy of Spenta Wadia. Anecdote connected with this photo is at http://arXiv.org/pdf/gr-qc/9705001.

ISBN-13 978-981-4299-57-2
ISBN-10 981-4299-57-X
ISBN-13 978-981-4299-58-9 (pbk)
ISBN-10 981-4299-58-8 (pbk)

Printed in Singapore.

DEDICATION

Lalitha Chandrasekhar

FOREWORD

In late 1980's, while I was working with Chandra on his biography, he mentioned a journal he kept of his scientific activities. One day he handed me a copy of what he had at the time. It was supposed to be confidential, and after his death, if his wife Lalitha and I thought it of interest and worthwhile, he would like to see it published. A brief glance at it evoked wonder and admiration. Chandra led a life of supreme and almost unparalleled effort in unraveling the laws of nature encoded in mathematics.

The year 2010 marks Chandra's birth centennial (October 19, 1910) and seems the perfect time to bring to light this unique document, *A Scientific Autobiography*. The journal gives a rare and personal insight into the joys and struggles of a brilliant scientist at work. For example, on his work on Radiative Transfer with which this journal begins, Chandra said, *"My research on radiative transfer gave me the most satisfaction. I worked on it for five years, and the subject, I felt, developed on its own initiative and momentum. Problems arose one by one, each more complex and difficult than the previous one, and they were solved. The whole subject attained elegance and a beauty which I do not find to the same degree in any of*

my other work. And when I finally wrote the book, Radiative Transfer, *I left the area entirely. Although I could think of several problems,
I did not want to spoil the coherence and beauty of the subject [with
further additions]. Furthermore, as the subject had developed, I also
had developed. It gave me for the first time a degree of self-assurance
and confidence in my scientific work because here was a situation
where I was not looking for problems. The subject, not easy by any
standards, seemed to evolve on its own.*"[a] It is this kind of insight
that illuminates the contextual circumstances surrounding his body
of work and gives it a depth of purpose we could not know otherwise.

As revealed in this scientific autobiography, and as Chandra
himself noted in the autobiographical account published with his
Nobel Lecture: "*After the early preparatory years, my scientific work
has followed a certain pattern motivated, principally, by a quest
after perspectives. In practice, this quest has consisted in my choos-
ing (after some trials and tribulations) a certain area which appears
amenable to cultivation and compatible with my taste, ability, and
temperament. And when after some years of study, I feel that I have
accumulated a sufficient body of knowledge and achieved a view of
my own, I have the urge to present my point of view* ab initio, *in a
coherent account with order, form, and structure.*"

This autobiography is a testimony to his having carried out his
quest to its perfection. The inner workings described in this document go beyond the vast landscape of physics, astrophysics and
applied mathematics. Chandra's published papers and monographs
evoke a feeling of respect and wonder. While to a casual student it
may seem intimidating and forbidding, for the serious-minded, however, they leave an indelible impression of their endearing value in
spite of the continual progress in the respective fields. They convince one of the innate values of science — the continuity, the inter-
dependence, and the necessity of combining original research with
scholarship. As we read, we find that with the perspective gained in
one area to his satisfaction, he leaves that area entirely and pro-

[a] *Chandra; A Biography of S. Chandrasekhar*, University of Chicago Press, 1991, p. 190.

ceeds to another with a complete sense of detachment, ready to start afresh in a new area. If it was necessary, he would attend classes, take notes, and studied as if he were once again a student. Or he would teach a course, and perhaps would give a series of lectures on the topic he wanted to learn. On the drop of a hat, he would fly to Oxford, England to have a discussion with Roger Penrose, or to Crete and Rome to work with his young collaborators, Basilis Xanthopoulos and Valeria Ferrari. In addition, this autobiographical journal reveals Chandra's human side, the man behind the legend — his intense association with his students and associates and his ability to inspire in others hard work and enthusiasm.

Chandra's extraordinary personality was characterized by an intensity and fervor for completeness, elegance, and above everything else, gaining a personal aesthetic perspective. It extended beyond his technical scientific publications to his semipopular lectures and essays. This is best illustrated, as he tells us in this autobiography, how he was led to writing the essay, *The Series Paintings of Claude Monet and the Landscape of General Relativity*, in which he speaks about the similarity of Monet's motivations in painting his "Series Paintings"[b] and his own motivations in the series of papers on black holes, colliding waves and scattering of gravitational radiation. In Monet's paintings, the same scene is depicted over and over again under different natural illumination and seasonal variations. The valley, the trees and the fields, and the haystacks are the same. Superficially, they may appear boring and repetitive. However, the different paintings radiate totally different aesthetic content. When seen as a group, a viewer can obtain a deeply convincing sense of the continuous nature of the experience in contrast to the shifting nature of what one observes. In a similar fashion, the same set of static symbols, which form the landscape of General Relativity, manifest in different roles in equations, unifying the description of vastly different physical phenomena, making the General Relativity as sometimes described as the one of the most "beautiful" theories.

[b] Chandra is referring to: (1) Haystacks (or Grainstacks), (2) Poplars, (3) Early morning on the Seine.

In concluding that essay, Chandra states he does not know if there has been any scientist who could have said what Monet did on one occasion

I would like to paint the way a bird sings.

But we do know of a scientist who spoke like a poet on one occasion

> *The pursuit of science has often been compared to the scaling of mountains, high and not so high. But who amongst us can hope, even in imagination, to scale the Everest and reach its summit when the sky is blue and the air is still, and in the stillness of the air survey the entire Himalayan range in the dazzling white of the snow stretching to infinity? None of us can hope for a comparable vision of Nature and the universe around us, but there is nothing mean or lowly in standing below the valley below and waiting for the sun to rise over Kanchenjunga.*

With such writings, often filled with parables, quotes from modern and ancient literature, with his Ryerson Lecture, *Shakespeare, Newton and Beethoven* and his book, *Truth and Beauty*, Chandra bridged the gap between what C. P. Snow calls the two cultures — the culture of sciences and of the humanities. This scientific biography, mainly concerned with an intense effort to understand Nature in the language of mathematics, may superficially seem too specialized and forbidding. But it should be no more forbidding than a memoir of a painter, who struggles with his tools, his trials and tribulations to attain his vision or of a great writer who creates characters and situations beyond one's ordinary imagination, bringing unexpected joy and insight.

In early summer of 1995, I had my last conversation with Chandra. I had received a complementary copy of his last book, *Newton's Principia for the Common Reader* directly from the publisher. I thanked him for it. He was annoyed that it took so long for it to be sent. He complained about getting weak, and his inability to do hard work and needed help to get back to his apartment after a walk. I reminded him, the days were extremely hot in Chicago at the time and that he should take care. I admonished him for working so hard. Take it easy, relax. "Yes, that is what I am doing," he said. "I am reading *Les Misérables* by Victor Hugo."

Chandra died on August 21, 1995 at the age of 84. The years I spent working with Chandra and writing his biography were the most enjoyable and the creative years of my life. As I sat with him in his office, among the books, journals, files of correspondence and sketches, conversing sometimes hours at a time, I often became transformed, and caught a glimpse of the incomparable world which Res Jost so aptly described:

> *There is a secret society whose activities transcend all limits of space and time, and Dr. Chandrasekhar is one of its members. It is the ideal community of geniuses who weave and compose the fabric of our culture.*[c]

It is indeed a privilege and honor to present for publication this scientific autobiography which evokes an enduring self-portrait of the man behind science.

Acknowledgements

I am grateful to Daniel Meyer and the staff at the Special Collections Research Center of the University of Chicago Library for providing me copies of the documents reproduced in this volume.

[c] Jost, a noted Swiss physicist, on the occasion of awarding Chandra, the Tamala Prize on January 9, 1984 in Zurich, Switzerland.

Thanks to Monona Wali for editing parts of this book and Heather Kirkpatrick for preparing the manuscript. Finally, I would also like to thank the Senior Editor, Lakshmi Narayanan and her colleagues at World Scientific for the excellent job in producing this volume.

Kameshwar C. Wali
October 2010

Preface

It is unlikely that I shall add any further instalments to this scientific autobiography that I have written at intervals during the past forty years. I am therefore, writing this preface in case there should arise at some future time, enough interest for its publication.

I have often described the years 1943 – 49 as the happiest in my scientific life. Those were the years when I was immersed in my investigations on radiative transfer and in the theory of stellar atmospheres: the years which culminated with the publication of my 'Radiative Transfer' and which included my exact solution of Rayleigh's problem of the illumination and the polarization of the sunlit sky and the unraveling of the continuous absorption coefficient of H^- and the continuous spectrum of the sun and the stars.

Half way during this period of intense and joyous activity, in September 1947, we went on a brief vacation to East Orleans, Cape Cod, at the invitation of Rupert Wildt. And while at Cape Cod, Lalitha repeated her suggestion of some months earlier that I write a 'history' of my series of papers on 'The Radiative Equilibrium of a Stellar Atmosphere.' Her suggestion arose from her having observed my total involvement in what I was then doing.

Before I left for Cape Cod, Paper XXI of the series had been sent to press, the end of the series (Paper XXIV was the last) was in sight, and the prospect of starting on my book was imminent. It was thus, in Cape Cod, in September 1947, that a preliminary draft of the first instalment was written; it was completed two years later when the manuscript of my Radiative Transfer was sent to press in September 1947. The second instalment was written in 1960 on the completion of my Hydrodynamic and Hydromagnetic Stability. And the last of them was written just about a year ago on the completion of my last book on 'The Mathematical Theory of Black Holes.'

The various instalments describe in detail the evolution of my scientific work during the past forty years and records each investigation, describing the doubts and the successes, the trials and the tribulations. And the parts my various associates and assistants played in the completion of the different investigations are detailed. But Lalitha, only rarely mentioned, was always present, always supportive, and always encouraging. And this is the place to record the depth of my indebtedness to her. But the full measure of it cannot really be recorded: it is too deep and too all pervasive. Let me then record very simply that Lalitha has been the principal motivating force and strength of my life. Her support has been constant, unwavering, and sustained. And it has been my mainstay during times of stress and discouragement. Thus, during the last months of 1981 when my last book was nearing completion, one snag after another kept on springing. Each had to be resolved patiently; but time was running short — very short. One such snag was particularly intractable. Lalitha identified herself so completely with my efforts that she said "there is pal hanging over everything; when will it lift". That is typical of her involvement in my work and of the way she has shared in my life: selfless, devoted, and ever patient and waiting. And so, I dedicate this autobiography, which is indeed my life, to her.

S. Chandrasekhar
January 22, 1983

CONTENTS

S. Chandrasekhar

A History of My Papers on
"Radiative Equilibrium" (1943–1948)

It all started with a letter from Placzek. He had just returned from England (1943 Fall). In a letter written from New York he asked me if any tabulation of the exact darkening function for the standard problem given in Hopf's tract exists. I had to admit that none existed but it recalled to me Gratton's paper (1937) in which he had applied the method of expansion in terms of spherical harmonics. And since I was lecturing on Radiative Transfer at this particular time, I went into Gratton's paper carefully and found that his methods could be improved upon. The calculations as given in Paper I were developed in these lectures. Also, for the sake of completeness, I had the Hopf-function tabulated. However, at this time, I did not realize that this was the first of a series of papers. The only other matter I can recall about this paper is that at Aberdeen (April 1944) H. N. Russell told me that he had read the paper with considerable interest (this was also the occasion he congratulated me on my election to R. S.). During this same visit to Aberdeen, I received from the *Mathematical Reviews* a paper by G. C. Wick on neutron diffusion for reviewing purposes. Reading this paper, I realized at once that the whole theory of radiative transfer as it existed at that time

I
Nov. '43

I
Nov. '43
June
July

II
March
–May
1944

could be done much better by adopting and generalizing Wick's idea of replacing the integrals which occur in the equations of transfer by Gaussian sums.

On my return to Yerkes late in April, I announced a colloquium on "Some recent papers on Mathematical Astronomy". The papers I reviewed were those of Thernoe on Emden-functions and of Wick. But for the colloquium, I thought it would be of greater interest to actually apply his idea to the standard problem. The calculations given in II (Secs. 1–5) were done during the last week of April and were presented at the colloquium. The calculations showed that the boundary temperature was predicted exactly on *all* approximations. When I gave the colloquium, I had not *proved* this. (The results of Sec. 6 came later.) However, in concluding this colloquium, I said that: "The clear superiority of this method over anything which has been done so far is apparent. Indeed, it is difficult to resist the temptation of redoing everything we know by the new method." I remember Henyey looking skeptical. But Sahade and Cesco were very enthusiastic: they came to my office after the lecture to say that they would like to work on some problem on the extension of the method. Mrs. Krogdahl and Miss Tuberg both came to my office later to say that they enjoyed the colloquium, and Miss Tuberg, in particular, expressed that she would like to do a thesis problem on the method.

However, the first thing I concentrated on was to establish the Hopf–Bronstein relation exactly. At first I did not see how it could be done. I remember that the crucial idea occurred to me after lunch. I had thought all morning and had not succeeded. But at lunch time, it occurred to me that if I understood why it was the *first* approximation that gave the exact result, then I could understand why the general nth approximation also gave the exact result. A little consideration (also during lunch) showed that this must result from a relation between the characteristic roots and the zeros of the Legendre polynomial (Eq. 68, II). Immediately after lunch, I verified that the relation could be easily established.

I did not realize then that in the course of the proof of the Hopf–Bronstein relation, I had eliminated the constants and expressed the solution in closed form. It is apparent *now* (Sept. 1947) that the extraordinary power of the method consists exactly in this: i.e. in the possibility of eliminating the explicit appearance of the constants of integration and expressing the solution in closed form. I was to realize this only a year and a half later (in Jan. 1946).

My immediate reaction to the whole idea was not one of doing anything beyond what was already known, but only to do better what was known: In other words, I did not draw the *obvious* moral from my proof of the Hopf–Bronstein relation. However, I did want to try the method out on more difficult problems. The solution for the Rayleigh phase function was something to try for — I recalled that in Milne's article in the Handbuch, he says that the problem with a phase function is too difficult to even consider formulating the equations! The method worked with surprising ease. And paper III was written and completed two weeks after Paper II had been sent in. Again, I did not realize that the solution can be found in closed form though all the essential steps are taken (Sec. 4). At about this time, Placzek visited us and I showed him my Calculations at the date. It was apparent to me at the time that Placzek was interested in similar problems in connection with his work at Montreal.

The next few months (June–July–August) were spent in routine problems. Sahade and Cesco were to work with me on the problem of absorption lines. They found the characteristic roots for this problem in the third approximation. Paper IV was sent in to press by the end of August. (I may add here that I have regretted this collaboration since Sahade and Cesco did not do their part of the work with due responsibility: I am therefore glad that this problem has *now* (Sept. 47) been exactly solved.)

After Paper IV I thought the problem was in spherical atmospheres. Here the problem was to find a means of replacing the derivatives in terms of the values at the Gaussian division. One evening coming to the Observatory, I was thinking how this should

III
May 31,
1944

May
1944

June
–July
–Aug.
1944

IV

Sept.
1944

be done when I met Sahade on the way and told him that the solution for the spherical-problem was the most important thing remaining to be done. I said that perhaps in solving the problem one should keep in mind the flux integral. This turned out to be the right idea: though I had thought fruitlessly for several days. By next evening, I had found the solution given in Paper V: I remember walking with Henyey to his house telling him how the problem had been solved. But even after this, the solution for the case $\kappa\zeta \propto r^{-n}$ in the second approximation was not too easy. I had to look to Watson quite a bit before the solution given in V (pp. 102–103) was found. This paper was completed in October, 1944.

Oct.
1944

After this time, I thought that the non-grey atmosphere was the one to concentrate on. One evening, Münch was in my office and I was telling him how important it was to bring up some sensible ideas for the theory of non-grey atmospheres. I showed him the sections in Unsold where the Rosseland-mean was derived. The looseness of the whole argument was apparent. Münch thought that the problem might be too difficult. I said, I did not see any reason why one should not go about it systematically: start with $\kappa_\nu = \text{constant} = \bar{\kappa}$. If κ_ν varies and is sensibly constant then the solution with a $\bar{\kappa}$ must be a first approximation. The question simply was "In what sense?" This will come from a perturbation theory. Thus talking to Münch, I developed all the principal ideas which are in my Paper VII. On the following day, I had worked out the $(2,2)$ approximation and established the new method of averaging κ_ν:

Nov.
–Dec.
1944

VII
Feb.
1945

$$\bar{\kappa} = \frac{1}{F}\int_0^\infty \kappa_\nu F_\nu^{(1)} d\nu\,.$$

But I did not at that time succeed in the general solution. A few days later, I was in Chicago, and while waiting in Zach's office, I played around and thought that the general solution should be found.

Dec.
1944

At about this time, Miss Tuberg was working on line formation problem with varying η. Her variational equations were similar to those which occurred in my problem. She had evaluated one determinant in the third approximation. The form of her solution immediately suggested to me that what was involved was the Van der Mondie determinant and the theory of symmetric functions. A reference to Perron gave the solution as given in Paper VII. The formal solution was completed by Christmas, 1944. But the numerical work delayed the completion to March 1946.

Feb.
1945
VI

During these months, Sahade and Cesco had done some more numerical work for the line formation problem. This was paper VI. Late in the fall of 1944, Kopal wrote to me expressing his interest in my series of papers and asked if I had considered the reflection problem and drew my attention to the inconsistency of Milne's solution with Hopf's relation. It was not difficult to work out the theory of the reflection effect on the new method — in fact, it did not take more than a few hours to complete the entire investigation. Paper VIII originated in this way. And thus also, my interest in the reflection problem. At about this time Kopal drew my attention to Ambarzumian's paper in the *Journal of Physics* and asked what I thought of it. Glancing through Ambarzumian's paper, I decided that I would return to the problem of diffuse reflection. But in the spring and summer of 1945, I had to turn to other matters. First, there was the article on Moving Atmospheres for the Bohr issue to get ready. And my frequent discussions with Wigner emphasized the H^- problem. Between April–October, 1945, I wrote the H^- papers I and II and the papers on expanding atmospheres, including the mathematical one to the *Cambridge Phil.* I returned to Radiative Transfer and Diffuse reflection only in November, 1945: i.e. after the Phil. Soc. Meeting in Philadelphia.

VIII
Feb.
1945

Nov.
1945

On returning from Philadelphia, it was my intention to go into the problem of diffuse reflection. But, as I thought, the matter was probably one of routine (i.e. to evaluate L_α's for various incident μ_0's: I was still not aware that the constants could be eliminated), I was not too enthusiastic about it. But an unexpected turn was introduced to my interests when, in a conversation with Kuiper, the accurate observations of Lyot on the polarization of the light reflected by Venus came to the front. The question then was what the theoretical expectations were. But to answer this question, one

IX
Dec.
1945

ought to have a theory of diffuse reflection. So I returned to the problem of diffuse reflection which I had sort of mentally shelved. Obviously, the first problem to tackle was that of diffuse reflection with the phase function $\lambda(1 + \alpha \cos \Theta)$. It was when working out this problem that it gradually dawned on me that the constants should be eliminated. At first I did not know the way to go about it, but eventually the technique became generally clear. The validity of the

Jan.
1946

reciprocity principle was a great help as a guiding rule. So Paper IX was finished early in January. And the relevant calculations in the second and third approximations were carried out. And the solution in closed form for the problem of Paper III was also found (actually somewhat later: see Appendix Paper IX). The importance of the H-functions was at last realized and the expression of the solutions in terms of these H-functions as a standard technique became clear. And the possibility of going to the infinite approximation was vaguely forming in my mind.

Once Paper IX was completed, I returned to the problem of

Jan.
–Feb.
1946

polarization. Meantime, I had become optimistic and had sort of boasted to Henyey about the success of the investigation as a foregone conclusion. But when I began to think of the problem, the difficulties of the problem became apparent. I referred back to the papers of Zanstra, Minnaert and Schuster. I also discussed with Herzberg. It seemed to me that there was nothing in the literature which was a sure guidance. The axially symmetric problem was therefore the first to concentrate on: because in this case, we know the plane of polarization and the intensities I_l and I_r in the meridian plane and at right angles to it were obviously the parameters to choose. It also became apparent at this time that the emergent radiation from an electron scattering atmosphere must be polarized: I had in fact, already talked to Struve about the possibilities of testing. Finally, in a weekend, I formulated the relevant equations and was surprised that they were reducible at all. I recall taking the Herzbergs to a movie that evening and explaining that I had solved the problem in essence. I was certain about the equations, because with $I_l = I_r$,

X   the formulae of Schuster were recovered. The paper (X) was written up for publication with the numerical solution of the emergent radiation in the second and the third approximations. The solution in closed form given in Sec. 6 came later. During the months following, the writing of Paper XI was very hard pressed for time. I had to go to Madison and later in April I was to give the lecture on "Scientist". But the importance of getting the solution of Paper X in closed form was essential before I embarked on the reflection problem. The solution was found enroute to Madison in the train.

At this stage (March 1946) there were two problems overhanging me. The first was the manner in which to parametrize polarized light and the second was to show that the H-function as I had defined it in terms of the Gaussian division and characteristic roots was related to the functional equation of Ambarzumian. From numerical evidence, it was clear to me that $H(\mu)$ as defined must tend to the solution of the functional equation

$$H(\mu) = 1 + \mu H(\mu) \int_0^1 \frac{\psi(\mu')}{\mu + \mu'} H(\mu') d\mu'\,.$$

I consulted Teller, Breit and Placzek regarding the first problem and von Neumann regarding the second: in both cases without any success.

The method of parametrizing polarized light finally became clear quite accidentally by a reference to Stokes' 1852 paper: in his collected papers. I recall that, disgusted with my fruitless efforts in getting advice from physicists, I collected, on a Saturday morning, all the books on optics in the library and saw what each of them had to say on polarization. And finally, when I came to Walker's Analytical Theory of Light (1904), I saw vaguely that the paper of Stokes, which is referred to here, was the paper to consult. However, it was only later in the afternoon that I went back to the Observatory and

May   brought Stokes' collected papers. The use of U became clear over the
1946   weekend and by Sunday the equations had been formulated. The az-
XI   imuth independent terms were unaffected and the solution already

found for these terms in April were therefore valid. Paper XI was completed on May 13.

After this, the principal problem was, therefore, that of relating my solutions for the diffuse reflection with Ambarzumian's. But before that, the problem of the general phase function had to be solved. This was done in Paper XII (completed in June). We went to Brown in July and after my return, I was first preoccupied with H^- (III) and the paper with Münch on the continuous spectra of stars.

Of course, all this time, I was worried about the H-functions. The basic theorem (Paper XIV, Theorem 1) dawned on me accidentally one evening when I was playing with it. I cannot quite recall how the idea came to me. I thought I did not know how to proceed for a long while. And suddenly one day I thought I knew! The method in the first instance was to assume that the theorem was true and construct a proof "going backwards."

The next question was to relate it to the Ambarzumian functional equations. I couldn't clearly see what Ambarzumian was about. So the best thing to do was to formulate them for the polarization problem. In this connection the lectures of Minnaert on the "Reciprocity principle" were of great value. His insistence for the case of polarized light, and my dislike for the particular formu-

lation of his, led me to consider the reflection of partially polarized light and the formulation of the reciprocity principle in terms of the symmetry of the scattering matrix $\vec{S}$. This paper turned out to be crucial for my investigation in Paper XIV.

With some difficulty, the functional equation for the scattering matrix $\vec{S}$ was formulated. And its reduction was a heart-breaking job. However, when I went to Princeton in September, I had derived the normal equations (XIV, 49–52) and the problem was to find ψ, ϕ, χ and ζ. The first step was to put my solution of papers XI and XIII in the form (41)–(44). This requires that $q^2 = 2(1-c^2)$. I verified this *numerically* in the Union Station between trains. All the time, in the train and in Princeton, I was worried I could not prove what I wanted, but the need for establishing the integral properties for

$H_l(\mu)$ and $H_r(\mu)$ became clear. Soon after my return from Princeton, the equivalences were established and it was a matter of only a week

XIV
Oct.
1946

or two till I had all the material ready for Paper XIV. This, together with XIII, were completed late in October. The principal problems I had in mind at this time are stated at the end of Paper XIV.

During November we had a whole train of visitors. Lindblad, Krishnan, then the Shajns. And I had the Gibbs lecture to prepare for. But the routine of getting the H-functions computed for the various problems was getting under way. For Rayleigh scattering, they were completed before the Gibbs lecture.

Jan.
1947

In early January, I wrote the Gibbs lecture and towards the end of January, I began to look further ahead ... (Lalitha was ill during this time).

Jan.–
Feb.
1947

According to the program outlined in Paper XIV (Sec. 20), the first problem was that of formulating the equations of transfer for elliptically polarized light. For this purpose I had to read Stokes paper more carefully than I had done hitherto. Once the paper was read, the solution was not far to seek. One weekend sufficed for Paper

XV
1947

XV. I was surprised that the matrix was reducible. At about this time, I began to think of the general laws of scattering and, having corresponded with Hamilton off and on during 1946, I decided to see

XVI
1947

him when I was due East in February. Paper XVI was, of course, simple. It was a tabulation of H-functions.

I realized at this stage that the main problems of semi-infinite atmospheres had been solved. It remained to go into finite atmospheres. For this purpose, functional equations similar to Ambarzu-

XVII
1947

mian's had to be formulated for the case of finite atmospheres. Moreover, the remarkable relationships between $I(0, \mu)$ and $S(\mu, \mu_0)$ was a source of considerable worry. The basic ideas underlying Section I (Paper XVII) also came during a weekend. At this time, I used to discuss a fair amount with Ledoux. I showed him the relation on a Monday morning. The functional equations derived in Section II (Paper XIV) did not take very much time. The main difficulty was the realization that *four equations* were involved and not two as Ambarzumian apparently implies.

All during the preceding year, I had been interested in getting more adequate tables of the Milne and Burkhardt integrals. These were getting completed at about this time.

Papers XV, XVI, XVII and XVIII were sent to press before I went to New York to give a lecture to Courant's group. At about this time, I realized that one should be able to express $H(\mu)$ as a complex integral. And I got into correspondence with Titchmarsh, which turned out to be very fortunate. The representation given in the Addendum to my Gibbs lecture arose from this correspondence.

Again in March, I was occupied with routine matters. But the
elimination of the constants was the first problem to tackle. And while I had thought of this on and off, I had not concentrated on it. But now this was the only outstanding problem. So, one weekend I sat down and formulated the basic problem. The basic mathematical problem as formulated in XXI (Sec. 3) was achieved during this weekend. I remember taking Ledoux to Walworth that Sunday morning and expressing my confidence that the problem could be solved. Later in the day, I thought I had it. I called up Henyey to tell him. But on Monday I realized that I had not in fact got it. However, I had accumulated enough confidence, and by Tuesday I knew how to get the solution. (At about this time, I started a correspondence with Davenport on this problem, who gave later an elegant determinantal solution.) With the interpolation problem $(F(x_\alpha) = \lambda_\alpha F(-x_\alpha)$, $\alpha = 1, \ldots, n, x_\alpha$ and λ_α assigned numbers) solved the expression of the emergent intensities in terms of the basic X and Y functions was immediate. At this stage, it became apparent that the X and Y functions must play the same role in the theory of finite atmospheres as the H-functions in the theory of semi-infinite atmospheres. Once the basic problem was solved, it remained to carry out the elimination of the constants for the various problems. During April and May this was accomplished. And in June I started writing the paper which was completed as planned on June 20. This was Paper XXI.

Paper XIX is a further tabulation of H-functions and Paper XX provides the exact solution for the standard line problem. The difficulty here was to pass to the limit

$$\text{limit}_{n\to\infty}\left(\sum_{\alpha=1}^{n}\frac{1}{k_\alpha}-\sum_{j=1}^{n}\mu_j\right)$$

The basic idea came from the fact that in the standard problem (Paper II); $Q = \sum_{\alpha=1}^{n-1} k_\alpha^{-1} - \sum_{j=1}^{n}\mu_j$ and this is $K(0)$. By analogy one would expect that $\sum_{\alpha=1}^{n} k_\alpha^{-1} - \sum_{j=1}^{n}\mu_j$ was related to the moment of $H(\mu)$.

Papers XIX, XX and XXI were sent to press in June 1947. The Compton scattering paper was written during July. The problem had, however, been solved in February during a weekend.

Sept.
1947
(Cape
Cod)

XXII
Sept.
–Oct.
1947
It remains *now* to complete the theory of finite atmospheres along the lines of Paper XIV. The basic theorem was established in July, but I have been too busy to start a frontal attack on the problem. That is the first thing I should do when I get back. But already during the vacation, I began to get worried about the integrodifferential equations for X and Y. However, I soon convinced myself that they are compatible with the functional equations for X and Y.

After we came back from Cape Cod, we had the Davenports. One of the first things I had to settle was the difference between Kestelman's formula and mine. Davenport located an error in Kestelman's formula. I am afraid that I could not get much other assistance from Davenport.

A week or so after the Davenports left, I began the reduction of the functional equations governing the laws of diffuse reflection and transmission to the X and Y equations. First, of course, certain integral properties had to be established. There was no difficulty in this. The reduction for the case $\lambda(1 + x\cos\theta)$ was easy enough. But the difficulty concerning $\lambda = 1$, which had haunted me for several months, began to loom larger. I thought I would get some clues by considering Rayleigh scattering. The reductions were not possible as it was apparent that in some way X and Y defined in terms of the reduced number of zeros should not be identified with the solutions of the functional equations. From analogy with the case $\varpi_0 = 1$, it seemed that wherever X and Y occurred, one should use

$$F_{(\mu)} = X_{(\mu)} + Q\mu[X_{(\mu)} + Y_{(\mu)}]$$

and

$$G_{(\mu)} = Y_{(\mu)} - Q\mu[X_{(\mu)} + Y_{(\mu)}].$$

Then it occurred to me to find out what the equations were which these functions F and G satisfied. And I found that they satisfied the same equations as X and Y! It was then that I realized that the solution of the equations for X and Y were not unique in conservative cases. And if not unique, we are entitled to use solutions for which $\int_0^1 Y_{(\mu)}\psi_{(\mu)}d\mu = 0$. The *standard* solutions were thus introduced in the theory. With the standard solution, the reduction for the Rayleigh function was possible on the understanding that the functions X and Y defined in terms of the reduced number of zeros were in the limit of infinite approximation to be associated with the standard solutions. (At this stage, I had guidance only from numerical work.) However, even so, in the end there was an ambiguity, as the two constants c_1 and c_2 introduced in the solution satisfied an equation of the form

$$(c_1 + c_2 + a)(c_1 - c_2 + b) = 0.$$

The question which of the factors was to be zero could not be decided. And for a while I even played with the idea of putting both factors equal to zero on the strength of the principle of "mathematical fairness". I finally decided that the Rayleigh scattering should provide the clue. Starting the reductions, the equations looked so impossible that I set $Q = 0$. (This was in some ways fortunate, for though Q turned out later to be an essential feature of the problem, the various conditions on the constants were not so complicated when $Q = 0$, as they later were found to be, that they could not be solved and at the same time required the essential tricks which were needed later.) The reductions with $Q = 0$ were carried through. The question which remained was: Is Q really zero? After having wasted a week, I decided that I would retain Q, and see what happens. This was the Sunday when Willie and Frederick came for tea. An error of sign seemed to indicate that Q was zero. I was very puzzled about this and I worried myself all evening, and, as we went to a movie with Bengt, I worried Bengt also. After we returned from the movie, I started working again and soon realized that Q, was not zero. And by next morning, the reductions were completed and Q was found to be arbitrary. During the afternoon, I wandered around the golf course and suddenly it occurred to me that the K-integral must be used. Immediately everything seemed to become clear, and coming to the Observatory, I saw Bengt, and said that the ambiguity was resolved. Later in the evening, the conservative isotropic case was completed. And next day, the differential equation for Q was also verified. The ambiguity for the other problems were similarly resolved. After October 15 and for the following three weeks, I wrote the paper and early in November, Paper XXII was completed.

Nov. 1947

During the four odd years I have worked on this series of papers, Frances has done all the numerical work: first in computing the various H-functions in the second, third or fourth approximations and then in the laborious job of iterating them to get the exact functions. Papers XVI, XVIII and XIX are published jointly with her. Her assistance in the numerical work has been of invaluable help to me. Indeed, the importance of such assistance for a theoretician has become very clear to me. For, the awareness that the numerical calculations will be done, gives the investigator freedom to think far ahead and on occasions actually use numerical evidence in guiding his work. For example, during the investigations of Paper XXII, it occurred to me that the X and Y functions defined in terms of the reduced number of zeros must be associated with the standard solutions, (i.e. $\int_0^1 Y_{(\mu)}\psi_{(\mu)}d\mu = 0$) of the functional equations. I had no reason to expect this as it was only a hunch. But its truth seemed inevitable because of the simplicity which would result. I did not want to spend my efforts proving this. I was worried far more with resolving the ambiguity in the solutions in conservative cases. So,

Frances computed (it took a solid week and more) some standard solutions and the results convinced me that I was along the right lines. The actual proof for this association of the X and Y functions defined in terms of the reduced number of zeros and the standard solutions in the limit of infinite approximation became clear only much later. But by that time, the entire problem had been solved.

After Paper XXII was completed, I could see that the main effort should, be put in the computation of the X and Y functions.

Nov.
–Dec.
1947
Frances computed several of these in the third and fourth approximations. But I began to feel that there was not much point computing these approximate solutions if they were not going to be iterated to obtain the exact solutions. And hand iterations seemed out of the question. So I went to New York late in November to the Watson Laboratory to see if they could not be persuaded to put the problem on the I.B.M. machines. Eckert was very cooperative and Mrs. Herrick was assigned to the problem.

Also, during this time I began contemplating a book on Radiative Transfer and began corresponding with Mott and the Oxford University Press. They were favorable to the idea and I agreed to have the manuscript ready by July 1948.

XXIII
Jan.
1948
During my absence in the East, Frances had computed the H-functions for the problem $\varpi_0(1 + x \cos \theta)$. (Altogether we had computed by now over forty H-functions) — these calculations formed Paper XXIII.

My plan had been to start on my book in January. But I found that there were a number of things I had to complete first. One of these was the solutions for X and Y which would be valid for

Feb.
–March
1948
small values of τ. The idea that something should be done in these directions came from discussions with van de Hulst. Van de Hulst had been trying to solve the planetary problem by considering the light which had been scattered once, twice etc., in the classical manner of King, and so on. But I told him that he could relate his problem to the standard problem I had been treating and that for the latter problem he should arrange his solution to be in conformity with

my exact solutions. If he did that, he would essentially be getting representations of my X and Y functions valid for small values of τ. This van de Hulst did and obtained, at the same time, an interesting interpretation of the X and Y functions. (I do not believe that this last is as significant as van de Hulst apparently thinks.) From the form of van de Hulst's solution, it was clear to me that what he was doing was essentially solving the X and Y equations by an iteration process starting with $X = l$ and $Y = e^{-\tau/\mu}$. The surprising thing is that van de Hulst did not realize this himself. In any case, once this is realized, the application of the idea to the general X and Y equations (van de Hulst had considered only the isotropic case) is immediate. The formal theory is almost trivial; and the only point of such a theory would be if it were accompanied by tables of the auxiliary functions. Computations were started on these. I wrote the paper up in February, but the calculations went on into April. Paper XXIV was finally sent in with an appendix (with Frances) on the F and G functions.

XXIV
April
1948

Already in March I had begun my book on Transfer. I hoped to finish it by August.

Now that Frances was leaving, I rather imagine that there will be no further papers in the series. What remains are the odds and ends: but the computations of these require a person who has experience. But once the X and Y functions come from New York, I should get my big paper on the "Illumination of the Sky" completed. I do not suppose that this will be ready before Xmas. I shall probably send it to the Royal Society for their Transactions.

The book was completed in September. Last proofs sent in November '49. Published February 2, 1950. Copy received February 2, 1950.

TURBULENCE; HYDROMAGNETISM
(1948–1960)

1947   At about the time my work on radiative equilibrium was coming to an end, and I was contemplating writing my Radiative Transfer, I began to think of the future. As I told van de Hulst, my work (and that of most other theoretical astrophysicists) during the two preceding decades was largely the carrying out of the program implicit in the pioneer investigations of Karl Schwarzschild, Eddington, Jeans and Milne. Clearly, they could not have foreseen the "results" of the subsequent investigations; indeed, they should have realized themselves that their own efforts would not lead them anywhere near the solution of the problems they had formulated. Still they had thought of the problems and showed the way for future developments; though their own efforts must, from a later vantage point, appear no more than skirmishes. And I asked van de Hulst, "When will we think of problems, new problems, problems for the first time, problems which will find their solutions only a decade or two later?" With these thoughts I had made up my mind that once I had written Radiative Transfer, I should leave the fields of my past endeavors and start on something entirely different. But I was not clear as to what that was going to be.

Fall 1948 After Radiative Transfer was finished in September 1948, we left for a short vacation in Bayfield; and while there, I decided that I would embark on turbulence: since turbulence is a phenomenon of the large scale and the essence of astrophysical and, indeed, also of geophysical phenomena is the scale.[d]

And so, on our return from Bayfield, I started my Monday evening seminars. The first audience included: Münch, Osterbrock, Edmonds, Huang, Brown and Code. At the same time, I lectured on Radiative Transfer on the campus. My class of two consisted only of Lee and Yang. The whole of my class of 1948 was to receive the Nobel Prize!

I started my seminars with Taylor's paper on "The diffusion by discontinuous movement"; then the paper by Karman and Howarth, and Batchelor's report on Kolmogoroff's theory. In the winter, the paper principally discussed was the one by Heisenberg. I found that for the stationary case, the solution of Heisenberg's equation can be found explicitly; also that in the case of decay, Heisenberg's integral equation, for the similarity solution, can be reduced to a differential equation which can be integrated. (The integration of these equations was the first job which Donna undertook, besides the polarization work, which was a "left-over".)

Spring 1949 In the spring of 1949, I spent a month at Princeton. At Princeton, I lectured on Radiative Transfer and on turbulence. Von Neumann came to the lectures on turbulence; and discussions with him disclosed certain errors in my judgment: thus, contrary to my belief at this time, the Heisenberg theory does not predict a cut off wavelength.

[d]The questions "Why is astronomy interesting; and what is the case for astronomy?" have intrigued me; I have often discussed these questions with my friends and associates. Granted that physical science, as a whole, is worth pursuing, the question is what the particular case for astronomy is? My own answer has been this: Physical science deals with the entire range of natural phenomena; and nature exhibits different patterns at different levels; and the patterns of the largest scales are those of astronomy. (Thus Jeans' criterion of gravitational instability is something which we cannot experience except when the scale is astronomical.) Of the many other answers to my questions, I find the following of Wigner most profound: "The study of laboratory physics can only tell us what the basic laws of nature are; only astronomy can tell us what the initial conditions for those laws are."

Jan. 1949

When in January I was asked to give the Russell Lecture at the meeting of the American Astronomical Society in Ottawa, I decided to give it on turbulence.

Later in Washington, D. C. I presented a paper on my work on Heisenberg's theory. This paper was sent to the Royal Society.

Aug. 1949

On returning from Ottawa and Washington, I concentrated on an essay on Stellar Structure for a volume on Astrophysics that Hynek was editing.

With the beginning of the new quarter, I renewed my seminars, with an account of Robertson's paper on "Invariant Theory". This led me to generalize the theory of isotropic turbulence to axisymmetric turbulence: this was a formal development which was useful in that it led me to cut my teeth into the subject. The long paper which came out of this was later published as a Transaction by the Royal Society. The question of gauge invariance was a troublesome one; and discussions with André Weil and Gregor Wentzel were very helpful.

Spring 1950

T. D. Lee had by this time joined me at Yerkes. And in the winter and spring, I turned my attention to developing an invariant theory of hydromagnetic turbulence. Again, interesting but only formal development.

Also during spring, I lectured on interstellar matter; and this led me to consider Ambarzumian's integral equation to describe the fluctuations in brightness of the Milky Way derived from an invariance principle. I extended Ambarzumian's discussion to the case of a finite atmosphere; and Münch found the solution for the infinite case. We thus started our series of joint papers; this was to keep me occupied during the following six months.

Summer & Fall 1950

I spent a few weeks during the summer at Ann Arbor lecturing on turbulence to their summer school. I also gave a colloquium on the polarization results. Uhlenbeck, who presided on the latter occasion, told me that Kramers had taught him about Stokes parameters.

The first complete results on neutral points in the polarization had been completed by this time; and an article (jointly with Donna) was sent to *Nature*.

Starting in the fall, I spoke on pressure fluctuations in my seminars. A generalization to hydromagnetics was straightforward and I wrote a short note for the Royal Society.

All this work on turbulence during 1950 was carried out when the affairs of the Observatory were "critical". Struve had resigned and Stromgren was to join us in January. And I was carrying out the difficult negotiations relative to the *Astrophysical Journal*. The latter was in so precarious a stage that the American Astronomical Society had washed their hands off the *Astrophysical Journal* and decided not to concern themselves with the journal. And since we were leaving for India in February, I left a day earlier and on a Sunday morning Spitzer, Schwarzschild, and I discussed the whole problem. I laid the case of the *Astrophysical Journal* to Lyman (who had been the chairman of the American Astronomical Society committee on publications: a committee which had been dissolved in December). And after explaining to him the case, I told him to frame a "constitution" considering fairly what I had told him. This he did during the afternoon; he showed me his draft "by-laws" at a party in the evening. I agreed that it would be "ok" with the University of Chicago — a strange agreement between two persons neither of whom had any authority. The final agreement which is now on the books is essentially the one to which Lyman and I had arrived at this meeting; and how it came to be is a different story.

We left for India on February 1, 1951. (In England, I was formally admitted to the Royal Society by Lord Adrian. Since I was the first Fellow whom Adrian admitted, a painting of this was later included in an exhibition by Brian Thomas illustrating "Ceremonies in London" for the Festival of Britain.)

On the return trip, I took stock of the efforts of the two preceding years; I was disappointed: a number of details had been clarified. But what was accomplished was not inspiring.

One fruitful idea which occurred to me during this same trip was that of adapting the ideas of the theory of turbulence to describe the fluctuations in density of the Milky Way. (The fifth paper of the series with Münch originated in this way.)

Jan. 1951

Feb. 1951

Spring
1951
On returning from England, I worked with Münch on the fifth paper of our series. Also, I gave a review paper on Turbulence to the Applied Mathematics Symposium of the American Mathematical Society.

In seminars during spring (when Ledoux was here on a visit), I developed the theory of density fluctuations and the effect of turbulence on Jeans' criterion. I also gave a sequence of seminars on shock waves and the expansion of a gas into a vacuum.

Summer
1951
During the summer, I wrote up my short paper on density fluctuations and the effect of turbulence on Jeans' criterion.

The feeling that I had acquired on my return flight from India that I was not really getting anywhere with my efforts in turbulence deepened. As a desperate attempt, I tried to incorporate into the theory of thermal instability the ideas of correlation functions, and in particular, the concepts of axisymmetric turbulence. The long paper in the *Philosophical Transactions* was the result. I recall that Osterbrock and Wayman were the two critical members of my seminars and helped much to clarify the ideas in this paper.

Fall 1951
With the beginning of the new quarter, I felt that I should perhaps change from turbulence to hydromagnetics. But I was not really getting anywhere; and by November, I almost felt like going back to some problems in stellar atmospheres and stellar dynamics. And while returning from the Observatory, and wondering what should I turn my attention to next, it suddenly occurred to me that I might generalize the classical investigations of Rayleigh and Jeffreys to hydromagnetics. And the moment this idea occurred, it was clear to me that a very large and fertile domain was open for study and investigation.

In my seminar the week following, I was able to present the entire basic theory of the inhibition of convection by a magnetic field. Osterbrock was clearly very excited about it. I called Fermi and made an appointment to show him what I had done. I was most encouraged by Fermi's reaction. In fact, I talked about it at the Institute seminar that afternoon and Fermi made some nice comments.

At about this time, Burbidge was visiting Yerkes and I recall telling him of the flood of problems which occurred to me at this time: the stability of jets, the effect of a magnetic field on Jeans' criterion, the Taylor problem (Couette flow).

December 1951 was in many ways a most happy one (in spite of a bout with mumps!). Not only had I recaptured, once again, an enthusiastic feeling for my work, I also received news of the award of the Bruce Medal.

Winter
1951
The winter months were busy ones. The paper on the inhibition of convection was completed and sent on January 15. (The complete solution for all three boundary conditions, the variational principle and the condition for overstability were all included in this paper.)

And in January in trying to fill out an open date in the colloquium schedule, I gave one on stellar scintillation largely based on a neglected paper of Rayleigh. The incorporation into the theory of ideas of turbulence was evident; and it was only a matter of a week before a paper on this was written up and sent to the Royal Astronomical Society.

Early in March we went to Berkeley and Pasadena. I gave an account of my magnetic inhibition work in the Bruce Lecture. I met G. I. Taylor at this time in Berkeley. And in Pasadena I lectured on Radiative Transfer and also on stability problems.

Spring
1952
It was in Pasadena, I first began seriously thinking of the Couette flow problem. (During this spring, Morgan resigned the Editorship of the *Astrophysical Journal* and I had to take it up.) Taylor's paper was clearly too "messy". I had, in fact, given seminars even before going to the West Coast; but the solution for the case when an axial field was present was obtained only subsequently: the crucial approximation that the values of the physical constants are such as to entail an important simplification had not occurred to me. Also the question of the boundary conditions was greatly puzzling. I discussed this with Wentzel; and indeed during a weekend he spent at Yerkes at this time, I discussed this matter with him an entire morning. We arrived at certain conclusions; but I later found that

the conclusions we had reached were untenable. So next Tuesday (I was a member of the University Council at that time) when I was on the campus, I had some 15 minutes to spare. So I called Fermi and went to consult him, confident that he would be able to resolve my difficulties. And indeed when I left his office 15 minutes later, I had with me the correct boundary conditions appropriate for perfectly conducting walls. With Fermi's boundary conditions, a variational principle was available. But the use of the principle required a "basis" for functions which together with their first derivatives vanished at the limits of the integration. I was not satisfied with my particular choice for the trial function; this matter of the basis was cleared up only years later (1957) but at this time I was too impatient to wait for such refinements. In any case, the calculations were long and Donna was occupied with these calculations during most of the summer.

Summer 1952

When I wrote the inhibition paper in January, I was, of course, aware that the extension to a spherical problem was of great interest. I wrote up the equations but soon had to give up on account of their complexity. When however the paper by Jeffreys and Bland came out, I realized that the simple spherical problem could be treated more simply and conveniently than Jeffreys and Bland had. Indeed, as a "stunt", Donna and I computed the characteristic values in the first approximation for the two cases (of a rigid and a free confining boundary) for $l = 1, \ldots, 15$ in exactly one day! The carrying out of the solutions in the second approximation took a few days more.

The paper on Hydromagnetic Couette Flow and on the thermal instability of a fluid sphere heated within were both completed and sent to the *Proceedings of the Royal Society* and the *Philosophical Magazine* (respectively) on the same day (August 29, 1952).

The paper by Jeffreys on spherical shells appeared at about the same time. To extend my method to the solution of this problem, the roots of $J_{-(\ell+\frac{1}{2})}(\lambda\eta)J_{\ell+\frac{1}{2}}(\eta) - J_{\ell+\frac{1}{2}}(\lambda\eta)J_{-(\ell+\frac{1}{2})}(\eta) = 0$ were needed. And Donna started on them.

In working out the theory for the spherical problem, I realized that errors involving scale transformations were easy to make: I therefore asked Garstang to check for them. He did not find any; but I am afraid he overlooked that an error had in fact been made. I detected this only a year later; and it was corrected.

Fall &
Winter
1952
In October it occurred to me that the effect of rotation on thermal instability (to the importance of which Jeffreys had drawn attention a long time ago) could be easily worked out. And I realized at the same time, that the question of overstability should be investigated. A variational principle was available and the work was completed by December.

During this period, I started my weekly meetings with Fermi to discuss hydromagnetic problems. These meetings originated at his instance. The first problem we considered included the effect of $\vec{H}$ on Jeans' criterion, the virial theorem, the gravitational instability of an infinite cylinder; and the nonspherical shape of magnetic stars. The weeks which followed are amongst the most exciting of my entire scientific career. Each Thursday, Fermi and I would discuss a number of problems; I would then work them out during the week; and the following week we would discuss what I had done; and discuss some further questions.

A small by-product was the effect of $\vec{\Omega}$ on Jeans' criterion. This was sent to Stratton's volume; but it did not come out for another two or three years (!).

In January, news came of the award of the Royal Astronomical Society Gold Medal.

Dec.
1952
There is one further incident which occurred at this time which I must record. At the Amhurst meeting of the AAS, Herzberg and I had a long talk on diverse matters. Among other things, he told me that he has been working on standards for wavelengths in the far ultraviolet. He had been disturbed by the fact that the theoretical value of Hylleraas for the lowest term of He II was below the experimental value. I told him that he need not worry as there was an error in Hylleraas' calculation. Herzberg was most surprised at this news.

He said that many spectroscopists had discussed the difficulty with Hylleraas' value at a meeting in Stockholm the preceding summer. And when I confirmed that I had detected an error in Hylleraas' in connection with my $\vec{H}$ work and while I had put it aside, I had no doubts about the existence of the error. Gerhard emphasized the importance of publishing my result. I said that since he had brought up the problem, we might work up correctly a twelve parameter function and publish a correct result. As I was to visit him in Ottawa later in the winter, I had promised to bring him the revised correct value. This was the beginning of the revision of the two-electron problem which had been considered as settled for 24 and more years!

During the winter months, I continued my weekly conferences with Fermi and by the end of March, our two papers had been completed.

Also the first paper on He II had been completed (jointly with Herzberg and Donna); and Herzberg made a return visit to examine how these calculations are made. (He wanted to extend the calculation to include more parameters.)

Spring 1953

In spring I spent a month at Haverford. Before going to Haverford, I had wanted to study the effect of a radial temperature gradient on the stability of viscous flow. The corresponding roots for the Bessel functions had to be computed; and Donna started on this, while trying to complete at the same time her long continuing calculations on polarization.

During spring, a number of odd things were cleared. First there was the paper on the polarization of the sunlit sky. The calculations had been completed some months earlier and during my absence in Haverford, Donna had typed the tables. The paper was written up, tables and all were sent to the American Philosophical Society for publication in one of their transactions in early July. Second, the companion paper (to the one already published in the Cambridge Philosophical Society) was also written up and sent for publication.

Once these things were gotten out of the way, the effect of H and Ω on thermal instability was the chief outstanding problem and Donna embarked on the calculations. The calculations are two parametric, more strictly three parametric. For given values of Q and T, $R(a)$ should be determined as a function of a, the minimum determined and the calculations repeated for another set of T and Q. In evaluating the minimum, things were not going smoothly; and during one of my absences, on her own initiative, Donna computed R for a whole range and discovered that the function has two minima whose relative magnitudes depended on R and T. The deeper of the two minima changes from one branch to another at a determinate Q; and a discontinuity in a occurs at this value of Q. Thus the discovery that for a given T, and increasing Q, there is a critical Q at which a changes from a large to a small value, was Donna's. (The question of overstability was not considered at this time; it was postponed for a later occasion.) The essential elements of the calculations were completed in time to be included in my Darwin Lecture.

While these calculations were going along, I collaborated with Nelson Limber on a short paper extending Ledoux's method of determining an approximate expression for the vibrational frequency from the virial theorem. Also at the same time, I worked out the effect of $\vec{H}$ and $\vec{\Omega}$ on Jeans' criterion. And in September, the paper on the effect of the radial temperature gradient on rotational instability was also completed.

With the beginning of fall, I wanted to start my seminars on something different. The paper by Rayleigh on the character of the equilibrium of an incompressible fluid of varying density seemed a good paper to discuss. The idea of extending it to include viscosity was immediately apparent. But I made a number of errors. I 'knew' something was incorrect in my treatment because in the case when the heavier liquid was overlying the lighter one, I found that the arrangement was predicted to be stable for wave numbers exceeding a certain value. I consulted Wentzel regarding this. He was very generous and checked my analysis. He first found that the basic equations were in error in not including the terms in $\partial \mu / \partial z$, I corrected for these terms, and the difficulty persisted. Wentzel then asked me if I had paid attention to the requirements that roots have a positive real part and ignored those that did not satisfy this requirement. I had not! Fortunately all this was clarified before we went to London in November for the Darwin Lecture.

My Darwin Lecture was on problems of hydrodynamic and hydromagnetic stability.

In England I gave several lectures principally on my stability work. We had to return from England on a particular date, namely, November 27 (which was a Friday); for I had agreed to give a lecture to a special symposium on Applied Mathematics (arranged by Peter Weyl) to be held at Northwestern University on Saturday morning, November 28. And the subject had been announced as "Characteristic value problems in high order differential equations which occur in problems of stability in hydrodynamics and hydromagnetics".

The plane which left London instead of coming directly went to Azores because of engine troubles, then to Gander, then to Boston, and then to New York. It was a trip during which I was completely sick. In order to get my mind off myself, I began to think seriously of how one could solve Taylor's problem in a simple yet in a systematic and an accurate way. And it suddenly occurred to me that the way to do it was to split the original differential equation of order six into one of order four and one of order two with four boundary conditions on the solution of the first equation, and with a choice of a basis which will satisfy the other two boundary conditions, the remaining second order differential equation then being used to derive a characteristic equation. Once the idea had occurred, it was obvious to me that the key to the solution had been found; and I felt that it must work. This made me so excited that the air trip became bearable ... With the delay in the time of arrival at New York, we could catch only a 6 p.m. plane for Chicago. We arrived in Chicago in time to catch a 10:30 p.m. train from the Union Station.

Dec. 1953

We finally arrived home well past midnight. I had to get up by six the next morning and drive to Northwestern University to give my promised lecture. At the lecture I presented the solution I had thought of on the trip as one whose success was assured and a foregone conclusion. I carried through the details of the calculation

Jan.
–April
1954

during the weekend and Donna started on the new method as one of high priority. It was all completed in one month. I sent the paper to Davenport for the first issue of his *Mathematika*. This paper presents a critical stage in the development of my ideas in the solution of stability problems. The symposium paper was also written during the same week.

With these papers out of the way, the remaining months of the winter and spring were essentially devoted to writing up and completing a number of loose investigations. First, there were the papers on the Rayleigh–Taylor instability — both the plane and the spherical problems. I should mention here that before going to England, I had found a variational principle for this problem: the first of the non-linear variety. Hide then used the principle to solve several of the remaining problems approximately. The paper on the plane problem was sent to the Cambridge Philosophical Society and the spherical problem was also completed soon thereafter and sent to Ferraro's Quarterly.

The paper on the effect of H and Ω (on thermal instability) was completed in April; and after this we went to Princeton for three weeks.

But before leaving for Princeton, I worked out the theory of the inhibition of convection by a magnetic field for the case when $\vec{H}$ and $\vec{g}$ are not parallel. In my 1952 paper, I stated that when $\vec{H}$ and $\vec{g}$ are not parallel, the component of $\vec{H}$ along $\vec{g}$ was all that was relevant; and that the onset of inhibition must be as rolls. I had asked von Neumann during my earlier visit to Princeton whether this was "obvious" to him. He said that it was not only not obvious to him, he felt on the contrary that the onset would not be as rolls. Martin Schwarzschild seemed to feel the same say. I had put this matter aside hoping to look into it at some later time. I now felt that the matter could not be postponed since the effect of $\vec{H}$ and $\vec{\Omega}$ were being considered and if one had to take into account the separate inclinations of $\vec{H}$ and $\vec{\Omega}$, the number of parameters in the problem would become unmanageable. The simplest context in which to clarify the situation is of course the case when $\vec{H}$ alone is present. Even when $\vec{H}$ alone is present and is inclined to $\vec{g}$, the determination of Rc depends on a minimization of R as a function of the two wave numbers a_x and a_y in two orthogonal directions in the horizontal plane. The variational principle (which continues to exist) makes the solution possible. But the calculations are inordinately long; and at one time I was not even certain whether I was not getting confused with using complex conjugates in a consistent manner. Wentzel examined my worksheets and assured me that I was not making any error ... Anyhow the formal theory for the two symmetric cases (bound-bound and free-free) had been completed before I went to Princeton, and Donna started on them during my absence. The bound-free case is quite a bit more unmanageable: the difficulty is in exhibiting the results in a manifestly real form; and the matter was completely resolved only after my return.

In any event, when we went to Princeton, it was clear that in the matter of thermal convection there remained only two outstanding problems. The first was to solve the overstable case in the presence of rotation for boundary conditions besides those for two free surfaces. This was important since the experiments of Fultz and Nakagawa with mercury were in disagreement[e] with the results derived from the Rayleigh boundary conditions. The second outstanding problem was to allow for overstability in the case when $\vec{H}$ and $\vec{\Omega}$ are simultaneously present. Both these problems were clearly heavily numerical. And while I foresaw months of calculations for Donna, there was not very much for me to do: the working out of the theory would require only a slight effort on my part. It was clear that I had to turn my attention to something besides stability.

[e]They had discovered overstability in accordance with prediction, in experiments with Hg. Their discovery had already been announced in *Physical Review*.

At Princeton I lectured principally on stability problems. I recall that Dyson and Goldberger were among those who attended my lectures. During the visit in Princeton, Tuckey asked me to talk to his study group on turbulence. And since at this time I was disappointed with the status of the subject, I gave a colloquium, moderately "frivolous" and cutting cruel jokes about the "superstitions" of the subject and the prevalent complacency in spite of the lack of any really rational theory. Martin Schwarzschild who was present at this colloquium was clearly irritated by my frivolity; and at the end of the talk, he told me on our way back to Prospect Avenue that he did not like my frivolity; and in any event what was I going to do about it! This was like throwing cold water on my face: and I began to think seriously once more about the subject; and I was to continue thinking about it during the rest of the spring and summer months.

Returning from Princeton, I concentrated on my second inhibition paper and by July all six papers which I had wanted to complete on my return from England had been written. (The theoretical prediction that when $\vec{H}$ is at an angle to $\vec{g}$, the onset of convection must be as rolls, was later to be confirmed by the experiments of Lehnert and Little.) Donna then started on the long calculations allowing for overstability in the case when Ω is present and the boundary conditions are general; and I started thinking of turbulence.

Summer
1954

With my renewed interest in turbulence, I began in July a series of seminars on the theory of turbulence. I gave a total of 14 seminars of which seven turned out to be on a new theory I was developing. It was during this time that it occurred to me (while driving to Chicago on a Thursday) that one might choose the set of moment equations by considering the correlations at two different instants at two different places and then using the quasi Gaussian approximation. I discussed these ideas with Fermi during lunch one day and he seemed quite interested in them. I felt at this time that I had made a real break in the theory of turbulence; but this elation and the hopes that it raised were soon to be dashed.

The question whether the defining scalar $Q(r, t - t') = Q(r, \tau)$ should be even or odd in τ was a very troublesome one. I even went to Madison to consult Wigner who was spending the summer there. Anyhow, the theory as I worked out and sent to the Royal Society in September had the wrong signs.

Fall 1954 In fall, starting in October, my Thursday afternoon lectures were on turbulence. Roberts and Prendergast were my regular audience; and George Backus was a member of the class.

Heisenberg visited Chicago during the fall. I showed him my paper and he seemed very satisfied with the results. He referred to it (as well as my earlier paper on his theory) in his lecture on the campus. These references made my lectures "fashionable" and my class increased in numbers. Savage was one of the new recruits.

My paper came back from the Royal Society with a referee's report which raised doubts about the choice of signs in my equations. This became a matter of acute discussion every Thursday with Roberts and Prendergast. I also discussed it with Wentzel. And I wrote to Doob also. Doob's reply, in particular, left no doubts that I was indeed wrong in my paper. The paper was finally corrected in January; and it is this corrected version that was printed.

Winter 1954 It was clear that if my theory of turbulence was along the right lines, then it could be extended to hydromagnetic turbulence as well. A problem which had occurred to me in 1949–50 (and about which I had talked to Fermi at that time) came to the fore: the problem concerns an elementary theory of hydromagnetic turbulence along the lines of Heisenberg's theory of hydrodynamic turbulence. This question occupied me during my seminars during the winter. The problem was finally resolved; and I was astonished that the theory gave two modes.

Herzberg who had meantime continued his calculations on the ground state of helium had obtained the results for an 18-parameter wave function. He was generous and included me as a joint author. This paper was communicated in February.

During the winter quarter, I gave my first set of lectures for the Physics Department: it was on Mathematical Physics (replacing Goldberger who had been scheduled). Trehan and Siciy were members of this class.

Spring
1955

At this same time, the experiments of Fultz and Nakagawa on the onset of thermal convection in mercury in rotation were coming to fruition; and Donna's calculations were also coming to an end. And in March the theoretical and the experimental papers were both sent to the Royal Society.

Somewhat later, Nakagawa carried out experiments on the magnetic inhibition of convection with Schein's magnet. I sent a preliminary note on his results to *Nature*.

I went to London in May (by MATS) to attend the Royal Society symposium on magneto-hydrodynamics. At this symposium, I presented my papers on hydromagnetic turbulence; and described also the first results on the magnetic inhibition of convection.

I spoke to the Institute sponsors in the spring; and an officer of the ONR who was present at this lecture evinced interest in the setting up of a hydromagnetic laboratory at the Fermi Institute. Allison was receptive to the idea; and Nakagawa was anxious to change his affiliation. And since I was particularly anxious to confirm the theoretical predictions on the onset of thermal instability in the case when $\vec{H}$ and $\vec{\Omega}$ were both present, I initiated steps towards setting up this laboratory.

In June there was the symposium at Ottawa which Herzberg had arranged in honor of Dirac. Dirac fell ill; but the symposium was held in his absence. I gave three lectures: two on stability and one on turbulence. Wentzel sat through these lectures; and when I apologized for the circumstance that required his sitting through them (when each of the subjects had been thrashed out in private with him) he remarked with characteristic generosity, "No! it was like hearing all of the Ring consecutively, after one had heard the different parts singly and out of sequence."

Summer
1955

The principal problem in stability to which I now turned my attention was to allow for overstability in case both $\vec{H}$ and $\vec{\Omega}$ are simultaneously present. The calculations are now much more complicated: since for each assigned Q and T the frequency of oscillation for every assigned wave number must also be determined. I knew that the calculations would be long and complicated. So this was to be a standby for Donna. Meantime, I was thinking of how to get Kolmogoroff's law out of my theory. Also I began a series of summer seminars on problems relating to the origin of cosmic rays. Don Wentzel was a summer assistant; and both Backus and Siciy were spending the summer at Yerkes at my invitation.

During this summer, I thought I had finally resolved the question of deriving Kolmogoroff's law from my theory. The necessary integrations were carried out and I had the matter settled in my own mind before leaving for Los Alamos[f] and Guanajato. At Los Alamos I spoke about my theory and Marshall Rosenbluth in particular seemed convinced and interested. And during my absence Donna started on the long calculations of overstability. (Unfortunately, the formulae were in error and the calculations had to be repeated: fortunately there was not too much time wasted on this account.)

Fall 1955

On returning from Mexico, I started on the accumulated Mathematical Reviews. Among the papers I had to review, there was one by Lüst and Schlüter on force-free magnetic fields. I immediately realized that they were solving on the machine, the solutions of the wave equation in five-dimensional Eucleidian space; and the solution can of course be found explicitly in terms of Bessel functions and no machine calculations are necessary. I thought that a breakthrough in axisymmetric problems had been accomplished. With Kevin Prendergast, a solution for magnetic stars was also found. And two short papers were sent to the *Proceedings of the National Academy of Sciences.*

---

[f]Ken Watson and Murph Goldberger were behind this. My clearance came through a week before I was to go to Mexico. At the conference I was to present Fermi's ideas on cosmic rays.

Left to myself, I should have exploited the new results. But the continued interference and meddling by Backus was a constant irritant. He would hear something either from me or from Kevin and would immediately start working on it himself. I am afraid that this particular form of aggressiveness from a student was extremely annoying; and it was to be a constant source of distraction during the next six months.

Winter 1956

I had to present at the 600th colloquium in February; and I chose the origin of the earth's magnetic field as my subject. I believed that the times of decay could be prolonged by internal motions. But this turned out to be a mistake: I would not have fallen into the trap but for the constant necessity of having to contend with Backus.

One nice result which nevertheless came out at this time was the proof of the stability of the equipartition solution. Wentzel and Goldberger checked my analysis as I could not quite believe the result myself.

The frustration of these months was due also to the fact that the Royal Society rejected my second paper on turbulence with a most discourteous referee's report. I withdrew the paper, but continued the correspondence with the referee.

The referee withdrew some of his more blatant remarks; but the whole incident was an unhappy interlude. I went specially to Washington to talk to von Neumann; and corresponded also with Heisenberg. The one happy recollection of this period is the sympathetic understanding which Wentzel showed. When Murph Goldberger asked me how I could write to the Royal Society with such restraint on the face of such insulting behavior, Wentzel quietly answered, "Chandra can afford to show such restraint".

In any event the paper was rewritten and sent to the *Physical Review* in February.

Spring 1956

In March the papers on the axisymmetric fields and motions and on the lengthening of the decay times by internal motions were completed. A large amount of calculations was carried out. However, on the whole, this period was about the most frustrating in my entire experience to date.

During the winter and spring months, I lectured on quantum mechanics. I had an enthusiastic class: the class applauded at the end of the courses. Apparently this was the second time it had happened in the Physics Department.

Fermi's class in Nuclear Physics had similarly been applauded in an earlier year.

In spring Siciy died. I had her and Backus for dinner on a Thursday evening. She went into surgery on Friday, and Tuesday she died. I was to go to Baltimore and Washington on Thursday: and on that morning I had to be one of her pallbearers. I liked Siciy; and it was a great pity she had to die.

Since I was to go to Los Alamos during the summer, I tried to finish the second paper on the effect of $\vec{H}$ and $\vec{\Omega}$ before that time. This was done; and the paper was sent in June.

Before I left for Los Alamos, I arranged for Donna to compute the various Bessel functions needed in axisymmetric problems in hydromagnetics. These calculations were published by her as a Supplement.

Summer 1956

During the summer in Los Alamos, I learned plasma physics; at the same time, I read Landau and Lifschitz's classical fields in preparation for my lectures on electrodynamics and optics during the following year. In plasma physics, I began with a study of the Boltzmann equation. (I should say that my excursion into plasma physics was not a wholly happy one.)

At Los Alamos, I arranged with Metropolis the integration of the equations of hydromagnetic turbulence by methods analogous to those I had used in my theory of turbulence.

One of the first things I did on returning from Los Alamos was to write a general account of my ideas on turbulence for the Centenary Volume of the *Journal of Madras University*. It gave me an opportunity to state my own case as clearly as I could.

John Sykes and Trehan joined my group at this time. And while I lectured on stability problems on Thursdays, I took up in my seminars the paper by Chew, Low and Goldberger and my own efforts with Watson and Kaufman.

It was at this time my association with Donnelly began. One Thursday afternoon he suddenly appeared at my office at the Institute and said he was interested to talk to me about stability problems in rotating fluids; and that Onsager had suggested that he talk to me. I made an appointment for the following week; but forgot about it. Donnelly appeared just as I was leaving; I apologized and suggested that if he wanted to see me he must take his chance and look in somewhat earlier. And he did. I was impressed by his persistence and patent enthusiasm.

Donnelly told me about his problem in He II; and I got sufficiently interested in it and agreed to collaborate with him on his problem. Our association which was to become a very profitable one was thus begun.

I really did not understand the real physics of He II. But given the macroscopic description, it was not difficult to work out the stability theory. This is what I did. The two papers which resulted were sent to the Royal Society in February 1957.

In December, news arrived of being awarded the Rumford Medal and premium. This was a source of some encouragement after a year of frustration. I prepared my Rumford lecture very carefully; and when I went to Boston to receive the award on March 13, I had the lecture all written up (which was later published in Daedalus). By accident, I found the following quotation from Virginia Wolff which expressed very accurately my attitude to my work of the past years. This quotation ends my Rumford Lecture.

> There is a square. There is an oblong. The players take the square and place it upon the oblong. They place it very accurately. They make a perfect dwelling place. The structure is now visible. What was inchoate is here stated. We are not so various

or so mean. We have made oblongs and stood them upon squares. This is our triumph. This is our consolation.

Another quotation which I have often quoted on lecturing on this problem is the following from Ruskin.

> Why did not Sir Joshua — or could not — or would not Sir Joshua — paint Madonnas?
> ... There is probably some strange weakness in the painter, and some fatal error in the age, when in thinking over the examples of their greatest work, for some type of culminating loveliness or veracity, we remember no expression either of religion or heroism, and instead of reverently naming a Madonna di San Sisto, can only whisper, modestly, "Mrs. Pelham feeding chickens."

The main problem on which Donna was concentrating after the completion of the helium papers, was the problem of the rotating sphere. And this was completed before I went to Boston and sent to the *Philosophical Magazine*. Though this paper was a first break into the subject, I was not altogether satisfied. The variational principle was discovered somewhat later and sent to the *Philosophical Magazine* as a companion paper.

My work with Donnelly had renewed my interest in the exact problem in the stability of viscous flow: i.e. allowing for the curvature and without making the small gap approximation. I had realized that my method in the *Mathematika* paper would not work. Clearly what was needed was an adequate base for the expansion of functions which together with its first derivatives vanish at two ends. I had encountered this problem already in 1952, and the problem recurred again. I was thinking about this problem one Sunday and had a definite idea. That evening we were at the movie theater in Elkhorn. Outside the theater after the movie, I ran into Reid (who had joined me in January); and I asked him to come home with us so that I could tell him about it. This method was a literal generalization of the base selected in my 1952 paper. Reid worked out the details of the method during the following days. The application of these functions to the Benard problem were not encouraging. However, I happened to call Zygmund in some other connection, and I told him about my problem. He said something about formulating a Sturm–Liouville problem; and it suddenly occurred to me that one must formulate a characteristic value problem in fourth order equations. And the suitable problems appropriate for Fourier-type and Fourier–Bessel type expansions are, of course, self evident. And on the morning we were leaving for Boston, I briefly explained the idea to Reid and asked him to work it out for the Benard problem to see how it worked. On my return, Reid had worked out the first approximation for the Benard problem; and the method was unquestionably the right one. I developed the corresponding functions for the Fourier–Bessel type expansion; and Donna evaluated the characteristic roots. Reid and I wrote up a joint paper for the *Proceedings of the National Academy of Sciences.*

Spring 1957

Once the suitable basis for expansion had been found, it was a simple matter to develop a method for solving the exact Taylor problem. But a large amount of numerical work was necessary. And this was Donna's next major job. Since I wanted the theory to be tested by experiments, I suggested to Dave Fultz that he construct a replica of Taylor's apparatus with an inner cylinder exactly one-half of the radius of the outer cylinder. And to encourage his experimental efforts, I made a bet with him as to who — he or Donna and I — would get there first.

Actually the number of matrix elements that had to be evaluated were very large; and I was doubtful that with other incomplete work on Donna's desk we should get there first. Actually Fultz did: he telephoned his results while I was at Los Alamos during the summer.

As I said earlier, while I was at Los Alamos during the preceding summer, I had arranged with Metropolis to integrate the equation of hydromagnetic turbulence. By spring, these integrations were at hand; and Donna carried out the large number of subsidiary calculations needed to exhibit the results. This paper was written and sent to the *Annals of Physics* in June.

Also I had written up the paper on the solution of the Boltzmann equation for dilute plasmas in strong magnetic fields. Most of the ideas in this paper are due to Ken Watson. But I smoothed the corners out and did the "secretarial" job on it! And early in July we went once again to Los Alamos.

The year 1956–57 in contrast to 1955–56 was somewhat less frustrating; and the award of the Rumford Medal to some extent wiped out the bad feeling of the preceding year.

Summer 1957

The summer at Los Alamos was uneventful. I studied some more plasma physics. The principal thing I did was to polish up the theory of the stability of the pinch which had been worked out by Watson. And with Kaufman I worked on our fourth paper on the Boltzmann equation.

Fall 1957

During the fall and winter, I gave a two-quarter course on plasma physics. These lectures were later to be written up and published by Trehan.

While lecturing on plasma physics, I got interested in the adiabatic invariance of the magnetic moment. I worked through the paper by Schluter and Hartweg; and it occurred to me that a general hierarchy of invariants could be deduced. I was very much interested in this; and it was the paper I presented at Lockheed Symposium in Stanford in December.

Winter 1958

All during the fall, Donna had been working on the wide-gap problem. As Fultz had gone to Europe, the completion of the experiments with the wide-gap Taylor apparatus had to be shelved for the time being; but Donnelly continued them as the junior (to become later, the senior!) partner. However, Donnelly also constructed a viscometer with a ratio of radii of one-half to test the theoretical results for the case $\eta = 1/2$. These experiments turned out to be extremely successful; and the theoretical and the experimental papers were sent to the Royal Society in March.

I knew at this time that I must soon embark on my book on *Hydrodynamic and Hydromagnetic Stability*. But a number of things had to be cleared first.

First, there was the long standing problem on the oscillations of a viscous globe. I had suggested this problem to Mrs. Fan and later to Edmonds; but they could not see their way towards the solution. With the experience I had gained meantime in the solution of spherical problems, it seemed to me that the problem could not be too difficult. The principal idea, that the problem be considered as an inhomogeneous equation for the pressure distribution, came to me at once. But in working out the theory I was hasty and careless with regard to the boundary conditions; and the paper as I communicated to the London Mathematical Society in February was erroneous. The referee saw that something was wrong; and I had discovered the error myself in the meantime. The error was corrected and the revised paper was completed and sent in May.

Second, acceding to my request, Herzberg together with Hart had obtained a 20-parameter wave function for H^-. And I wanted to revise my calculations of twelve years earlier using this new function. This meant evaluating a very large number of new matrix elements. I evaluated them during two or three weekends; I was glad that Bill Reid checked them all. I set up the formulae and Donna first computed the corrections with the plane wave approximation for the free state. This paper was completed in April. The next thing was to evaluate the weight functions so that the Hartree functions could be used. The formulae were set up during a week I was in Chicago at this time; and Donna completed them during the remaining months; and as a joint paper this was sent in to the *Astrophysical Journal* in June.

Third, I had promised the thermal panel of the Defense Department to work out the theory of the diffuse reflection of a pencil of radiation by a plane-parallel atmosphere. In particular, I wanted to set up the integral equation from the principles of invariance. This I did in a short paper I sent to the *Proceedings of the National Academy of Sciences* in June. I gave an account of this paper at an Institute seminar where besides Urey and Reid only two or three others were present. Since no one knew of my work on Radiative Transfer, I spoke about the whole field in a general way for the entire hour and a half. And Bill Reid remarked to me later, "It is a pity Chandra that none of the astronomy students have a chance to hear you on these subjects."

Fourth, the time had come to send an article I had promised for the Planck Festschrift. Since I had wanted to clear up the thermodynamic significance of the variational principles in thermal stability problems for a long time, and since the principles must be clarified in my book, I thought a good deal about this. The old paper by Jeffreys was too special and too inelegant. Moreover, I was anxious to clarify the principle in case of overstability. This later puzzled me greatly; but the solution given in the paper occurred to me while I was talking about it in my seminar. This paper was sent in June.

Fifth, I had promised Anderson and Segre to write up an introduction to Fermi's papers with astrophysical interest. This was a difficult assignment: but it could not be postponed.

And finally, I was interested in the development of variational principles dealing with hydromagnetic equilibrium configurations. Woltjer was devoting considerable attention to this; but I am afraid that with this many distractions, my part in it was unsatisfactory and inconclusive.

Before I left for La Jolla for the summer, I arranged for Donna to extend the calculations in my 1953 *Mathematika* paper.

Summer 1958   The summer in La Jolla was an unproductive one except for one thing. Rosenbluth wanted me to look into the rotating pinch. And I discovered the variational principle which exists for these problems when one considers k^2 as the characteristic value parameter instead of w^2. On our return from La Jolla, I wrote out this principle for the simplest hydrodynamical problem for the Jubilee number of the Indian Mathematical Society.

Also during my stay at La Jolla, Allan Kaufman and I corrected an oversight in our Paper III: it was added as an Appendix to our paper IV.

Fall 1958 The fall was full of distractions. Lalitha went to India. I had to go to Los Alamos twice; once for a hydrodynamical symposium and once in connection with the Sherwood Committee. Then in December, I was to give the same lecture at six of the campuses of the University of California. And moreover, there was the 700th colloquium in December; I had decided to present on general relativity which meant a good deal of additional study.

The only "profitable" thing which happened during these months was during my lecture at Berkeley. Friedman was present at my lecture and he suggested what seemed a considerable improvement of the technique of my *Mathematika* paper. I worked this method out; and Donna's calculations showed that the new method did not provide any substantial improvement over my earlier method; in fact, none at all.

Winter 1959 With the New Year, the beginning of H.H.S. (H.H.S. stands for the book *Hydrodynamic and Hydromagnetic Stability*) could not long be postponed. But I was lecturing on Quantum Mechanics during the spring and fall; and there were still a number of distractions. Dirac visited in March; and moreover, the question of the asymptotic relation, $T_c = C(1 - \mu)^4$ as $(1 - \mu) \to \infty$ which was apparently valid for the Taylor problem was worrying me. Several attempts, which seemed promising, were tried; and two or three months were lost (both my time and Donna's time) in following deceptive trails.

Spring 1959 Finally in the spring, I started on my book in earnest. And the first chapter was written in April. And for another year the book was to occupy me.

Before I get into the details, I should say one thing. I wanted the book to have a certain logical structure with symmetry and pattern. And this requirement of a pattern forced lines of investigation which would not be suspected by a casual reader of the book. Also as it turned out, almost all of the numerical work incorporated in the published papers had to be redone differently; and a large amount of new material had to be worked out and assembled. In all this, besides Donna's constant effort and attention, I had also the assistance particularly of Reid and Vandervoort.

Chapter
II
May
–June

In writing Chapter II the principal questions which had to be clarified were the Boussinesq approximation, the proof of the extremal principles, the cell patterns, and a description of the experiments. The calculation of the streamlines for the hexagonal pattern alone took more than four hours of plotting and drawing not to mention the hours spent on calculation and interpolation.

Chapter
III
July

In writing this chapter I had to clarify my ideas on the vorticity theorems, the Taylor–Proudman theorem, and the propagation of waves in a rotating fluid. The calculations of my original paper (I) had to be recomputed by using a more systematic treatment based on a symmetric characteristic matrix. The question of overstability had to be clarified: an error (and a misstatement) in Veronis's paper was found in this connection. The results of Goroff on the efficiency of convection in the overstable regime were most timely.

Chapter
IV
July
–August

The introduction to hydromagnetics which this chapter includes required careful selection. The matter of the boundary conditions and its peculiar role in the meaning of the variational principle and the establishment of a new invariant were all matters which required attention and care.

Chapter V was a straightforward one.

Chapter
VI
Aug.
–Sept.

Chapter VI required a good deal of redoing. The new treatment based on the toroidal and poioidal functions resulted in a drastic rearrangement of the material. The problem of the onset of instability in spherical shells was completed and clarified. Lyttkens sent his results in time and with unexpected promptness. And the variational principle and its thermodynamic significance had to be treated far more generally than hitherto.

The first six chapters were completely typed by October.

Chapter VII on Couette flow required a large amount of additional work: the modes of vibration of a rotating column fluid; the stability of non-viscous Couette flow; the perturbation theory for the Taylor problem for $1 - \mu \to 0$, an analysis of alternative methods; the question of over stability; and finally an account of all the available experiments. And, of course, the cell patterns.

Chapter VIII led to a host of new problems: the real meaning of DiPrima's peculiar results (this became clear when I analyzed the regions of the fluid which were unstable according to Rayleigh's criterion and explained it to Reid during one of his visits); the question of spiral flow (this is the Goldstein problem which I had wanted to examine since 1952). It was clear that Goldstein's analysis of the latter problem was inadequate; but I was unprepared to find that he was altogether wrong. Both the non-viscous and the viscous flows had to be discussed; and, I arranged for Fultz and Donnelly to verify my theoretical calculations as promptly as possible for inclusion in the book. Two short papers on these subjects were sent to the National Academy in November and December. With all these new investigations, Chapter VIII could not be finished before the end of November.

Starting Chapter IX, I realized that the patterns of the early chapters required that I discuss the non-dissipative Couette flow first. Several new results had to be worked out; in particular I realized that a magnetic field of a certain strength could stabilize all adverse flows; and that some of Reid's calculations on non-viscous hydrodynamic flows could be used to determine the exact numerical constants. A feverish exchange of correspondence with Reid had to be carried out; and two short notes for the National Academy resulted.

Turning to the dissipative flows, the general case had to be treated; and Niblett's calculations for insulating walls had to be repeated. My original method of solution was clearly unsatisfactory; and the theory was worked out anew by using the C- and the S-functions tabulated by Harris and Reid. The general case of negative μ was also treated; results for $\mu = -1$ were to be included and finally the asymptotic forms for $Q \to \infty$ had to be discussed and the relevant equations investigated. And finally there was the problem of the curved channel; and Lebovitz undertook to do this.

On the experimental side, it was satisfactory that Donnelly and Ozima successfully concluded their experimental work in time for inclusion in the book.

Chapter
X
Jan.
1960
Again in writing on the Rayleigh–Taylor stability, the effects of rotation and magnetic field had to be included as well as surface tension. I found that surface tension could be included in the equations of motion as delta functions; and that the treatment can be unified.

In treating the effect of the magnetic field when it is in the direction of g, I had to contend with a novelty: the absence of a dispersion relation in the stable case. The clarification of this involved subtle questions. My first impressions turned out to be erroneous; this became clear during a discussion with Dalitz. My final resolution of the paradox in terms of the reflection and refraction of the Alfven waves was as surprising as it was simple. In the unstable case, in contrast, there is a definite dispersion relation; and this had to be evaluated. Again, in treating the problem in case $\vec{H}$ was at right angles to $\vec{g}$, I discovered that the effects could be interpreted in terms of an anisotropic surface tension: a useful concept when I came to treat the Helmholtz–Kelvin instability.

Chapter XI also included an account of the oscillations of viscous globe and of a viscous drop.

Chapter
XI
Feb.
1960
Chapter XI on the Kelvin-Helmholtz instability raised many new problems. The effect of rotation was particularly troublesome. A non-algebraic characteristic equation had to be considered. Discussions with Kaplansky and David Shafer were most useful. Actually the matter could not be resolved at the time; I had to leave it open. The problem was finally resolved in the pressure of the last weeks; a chance remark of Kaplansky's during an earlier discussion had the essential clue. I asked Vandervoort to do some numerical work on

the various branches of the solution; and on the basis of this work the problem was finally clarified.

In treating the effect of a horizontal magnetic field on the Helmholtz instability, an error vitiated the early arguments and delayed the chapter for several days: the matter was finally cleared up only during the last days!

Chapter
XII
March

In the chapter on the stability of jets and cylinders, the inclusion of the effect of viscosity in the capillary as well as the gravitational instability was a major undertaking. A large amount of miscellaneous calculations had to be done by Donna on a basis of high priority.

In treating the effect of fluid motions, I was quite annoyed by Trehan's clumsy treatment of the problem: he had effectively concealed all the symmetries and I had to do the whole thing *de novo*. The stability of the pinch had also to be included. With all these many new investigations, the chapter could not be finished before the last days of March.

Only three weeks were now left. (Fortunately, my Weizmann Lectures had been postponed by a week.)

Chapter
XIII
April

Starting Chapter XIII under extreme pressure, I realized that the virial theorem should have to be formulated in tensor form. The existing treatments had many loopholes and were quite unsatisfactory. I developed a whole new approach in discussing the problem one morning with Nelson Limber. I was quite pleased that the conservation of angular momentum is a consequence of the virial theorem; and that the non-spherical nature of magnetic stars was an immediate corollary. My earlier treatment of the pulsation problem with Limber had to be generalized as well. But it was an exciting week in spite of the extreme pressure under which I was working. (I later wrote up this work as a paper for Bellman's journal.) The rest of Chapter XIII deals with Jeans' criterion and in particular the effects of rotation and magnetic field. Fortunately Vandervoort had done all the necessary numerical work. It was about April 7 or 8 when this chapter was completed.

Without starting on the last Chapter XIV, I had to organize all the figures. (Those for the first six chapters had been sent earlier. I should have said that during all this time, the illustrations were being drawn and kept more or less up to date.) When all this was finished, I was so tired that I decided to go to New York to give my invited talk to the American Mathematical Society.[g] On returning from New York, the weekend and Monday were spent on the various sections of the book which had been left incomplete: the experimental sections in Chapters VIII and IX, the revisions of Chapters X and XI and others.

It was finally on Tuesday morning that I started on Chapter XIV. This chapter was to include the variational principles as the beginning of a "new story". Freeman *et al.*'s principle was to be the main theme. But his paper was depressing reading: it was so complicated and so clumsy, that I actually thought that I would abandon the idea of having a Chapter XIV. I knew this would disappoint Donna and so I decided that I would start on the chapter anyway and write out the principle for the incompressible case which I knew I could develop *ab initio*. Once I had written it, I could see how the principle could be generalized to allow for compressibility without having to read Freeman *et al.*'s paper. The theory was fully worked out by late Wednesday evening; and I wrote up a first draft before going to bed.

Early on Thursday morning, I started my second draft. By noon I was ready for the nth draft. (By this time, I was in a constant state of nausea.) But when I had come to the end by 5:00 p.m., I was not satisfied with what I had written. So I had to start all over

---

[g]This trip was an altogether nightmarish one. I had a bad cold and some temperature. And Russell Donnelly insisted on my redrafting his letter to the *Physical Review* on his results on the hydromagnetic Couette flow in the hotel in New York. Next day when I got onto the platform for my lecture, I felt so parched that I insisted on having a glass of water: to the consternation of the Chairman. But several people (apparently noting my sickness) dashed out to get me some water. I was shivering in the plane all the way back; and on arriving midway, I decided nevertheless to drive back home. On arrival I found I had a temperature of 103. Next morning I got up early and refused to have my temperature taken since I had to work anyway.

again; and it was finally completed by 9:30 p.m. I called Donna at that time and she came over to start typing the last chapter.

Most of Friday was occupied in filling in the formulae for Chapter XIV, and various minor details. (To make matters worse, Mrs. Schofield had been called away; and I had to make arrangements for the *Astrophysical Journal* in my absence.) Anyway by about 5:00 p.m., the manuscript was complete and ready; and I could at last start thinking about my Israel trip and the Weizmann Lectures.

Early on Saturday morning, Norman Lebovitz drove us to O'Hare. And at Idlewild, we had two most enjoyable hours with Martin and Barbara before we took off for England and Israel.

In London the following day, April 24, the manuscript was handed over to Mr. Wood of the Clarendon Press.

The Development of the Virial
Method and Ellipsoidal Figures of
Equilibrium (1960–1970)

1960   It all began while writing Chapter XIII of *Hydrodynamic and Hydro-magnetic Stability* in March of 1960. This chapter was to include the formula which Limber and I had derived for the frequency of "radial pulsation" of a magnetic star. But since Fermi and I had shown that magnetic stars are unlikely to be spherical, the concept of "radial oscillations" is untenable. I recalled the earlier papers of Rayleigh and Parker on the tensor virial theorem; and it seemed to me that what Ledoux had accomplished for radial pulsations of a spherical star with the aid of the scalar virial theorem, should be carried out for magnetic stars, with the aid of the tensor virial theorem. But on examination, it appeared that the papers of Rayleigh and Parker did not provide an adequate base. And attempting to think through the matter *ab initio*, I realized the importance of defining the tensor potential v_{ij} and the associated potential-energy tensor ω_{ij}. The key formula

$$\partial\omega_{ij} = -\int_v \rho(\xi)_k \frac{\partial v_{ij}}{\partial x_k} d\vec{x},$$

as well as the generalization of the formula for σ^i, which Limber and I had derived, are presented in *Hydrodynamic and Hydromagnetic Stability*.

The basic ideas underlying these developments came to me one morning when walking to the observatory along the golf course. I called Nelson on reaching the observatory and developed the entire theory in explaining the ideas to him.

I realized already at this time that the tensor virial theorem must have wide applications to the study of rotating and magnetic stars. But *Hydrodynamic and Hydromagnetic Stability* had to be finished; and a year was to go by before I seriously returned to the subject. However, I wrote up these parts (of the book) as a paper (1) for Bellman's journal; and it was published in the same year.

I remember discussing the theorem with Woltjer later that fall and telling him how I was planning to use it for a systematic study of the classical ellipsoidal figures. It seemed to me that Woltjer was entertaining similar ideas; but upon hearing of my own plans he apparently abandoned them.

An unforeseen event (crucial for the further developments) happened that Christmas. Norman Lebovitz, who was my student at that time, had been working on the dynamo problem. But it was getting nowhere. So I suggested to him that he consider applying my tensor-virial method to the problem of oscillations of the Maclaurin spheroid. My idea at that time was that the problem could be solved approximately by assuming a linear form for the Lagrangian displacement.

At first Norman was reluctant to embark on a new thesis subject, but with hardly six months left to complete his thesis he had no choice. So he undertook to investigate the problem; and soon, he became very enthusiastic.

Spring 1961

During the winter I was extremely busy with the proofs of *Hydrodynamic and Hydromagnetic Stability*. But occasionally Norman would report to me on his progress. I recall my slight annoyance that contrary to my suggestion that he make a linear assumption for $\vec{\xi}$, he was attempting to solve the problem exactly. He nevertheless persisted and his successful exact solution of the problem was beyond anything I had foreseen.

From the present vantage point, the three things that Norman's thesis accomplished were the following: (1) to show that the problem can be solved exactly, (2) that the evaluation of $\mathcal{D}_i$ (on which v_{ij} depended) was not "elementary," and (3) that the virial equations were required to be supplemented by a "solenoidal condition."

Norman's success convinced me that the application of the tensor virial theorem should be pushed with vigor. Thus, my plan of 1960 that I pursue general relativity with an undivided mind was to be frustrated. It did not seem so at the time; but the classical ellipsoids were to absorb a large part of my efforts in the seven following years. But after 1963, it was carried out largely in protest. I felt an obligation for the subject. Perhaps it was misplaced. But I lost some crucial years.

And when Norman left for M.I.T. after taking his Ph.D. degree in June, we planned to meet in Maine (where I was to spend part of June and most of July) and discuss the course of future developments.

June 1961

But one thing occurred to me earlier: Is there a generalization of Emden's formula for ω to rotating stars? It was a simple matter to derive the generalization and I wrote a "Letter" to myself. "A theorem on rotating polytropes" (2) was published that summer. I made some additions to it while I was in Maine. (The original Letter contained only the formula for ω; in the revised (published) version, formulae for ω_{11}, ω_{22}, and ω_{33} are also given.)

In July, Norman visited us in Maine; and we decided that the first problem to be attempted was to develop a general theory of the oscillations of rotating gaseous masses based on the tensor-virial theorem.

August 1961

On returning to Yerkes in August, I found some papers from Norman awaiting me. In these papers, Norman had undertaken to reduce the supermatrix and derived some basic formulae. But the developments could be much simplified by considering the superpotential χ. And August was a hectic month. The first paper on the superpotentials in the theory of Newtonian gravitation (3) and the paper "On the oscillations and the stability of rotating gaseous masses" (4) were both completed that month. In many ways, it was also an exciting month: seminars almost every day with frequent telephone calls to Norman at M.I.T.

We were to leave for India on September 2. And Norman drove from Boston to meet us at the airport in New York; and we spent several hours at the airport going over the final manuscripts of the two papers. And thus our collaboration during the next few years began ... Earlier in the same week, the first copy of *Hydrodynamic and Hydromagnetic Stability* had reached me. And with the book in my hand and the two papers, full of promise, completed, I was content when the plane took off for India that evening. Thirty-six hours later, I was in a new world.

We were in India during the four months, September–December. During those four months we traveled widely — Bombay, Calcutta, Kharagpur, Jamshedpur, Ahmedabad, Delhi, Hyderabad, Madras, Bangalore, and Kodaikanal; and I gave as many as seventy lectures. And while we were in Delhi we were invited to dinner by Nehru; and the evening was a most memorable one.

The Letter (5) on the interpretation of the double periods of the Beta Canis Majoris stars was principally Norman's contribution. The Letter was sent to the *Astrophysical Journal* during my stay at the Statistical Institute in Calcutta.

We returned to the United States in January 1962. And during the winter and the spring quarters I gave a first course in *Applied Mathematics* — and these lectures were to take a substantial amount of time.

And the first research item on the "agenda" was to write out my second paper on the stability of viscous flow between rotating cylinders using the adjoint equation. And meantime, Norman had sent his worksheets on the application of virial method to the os-

cillations of rotating polytropes. Norman had clearly been working steadily during my absence. I worked through the sheets and was able to simplify some of it. But I realized that the first thing I had to do was to learn the details of the derivation of the expressions giving the potential of an ellipsoid. I did not find Kellogg to my taste. I found Ramsay and Routh much more satisfactory; and I started working through Routh. And soon I discovered to my astonishment that the potential

$$\mathcal{D}_i = G \int_v \frac{\rho x_i'}{|\vec{x} - \vec{x}'|} d\vec{x}'$$

which had given considerable difficulty to Norman, was derived in Routh by an entirely elementary method. Not only $\mathcal{D}_i$, but also the potential due to any polynomial distribution in an ellipsoid could be found by the same elementary methods (due originally to Ferrer). This discovery of Ferrer's potentials was to be basic to all subsequent developments. I called Norman and told him about what I had found. This was the origin of the paper on "The potentials and the superpotentials of homogeneous ellipsoids" (7). In this paper, the notation was standardized and the systematic use of the A-symbols was originated. The second paper (6) on the superpotentials was a logical antecedent.

During this time (February–March) the calculations pertaining to the oscillations of a rotating polytrope were proceeding. But what was uppermost in my mind was to find some means of locating the point of bifurcation along the Jacobian sequence. So constantly was this in my mind that it almost became an obsession. Then on a Friday in May, it suddenly occurred to me that the second-order virial equation could not in principle determine the desired point of bifurcation; and that one must go to the higher order virial equations. And this fact became further apparent in a telephone conversation with Norman.

Spring 1962

I developed the third-order virial equations over the weekend. And I was able to show how the point of bifurcation can be located by using one of the integral properties provided by the virial theorem. Lynden–Bell arrived at Yerkes during this same weekend. And in a seminar on Monday, May 24, I was able to "announce" the entire development together with the numerical determination of the neutral point. Lynden-Bell commented after the seminar that my method appeared to him "as a triumph of notation".

The discovery of the usefulness of the third-order virial equation was one of the few moments of real exhilaration in this entire barren period. However, at this point my understanding of the usefulness of the third-order virial equations in these contexts was incomplete and some of my earlier ideas were actually erroneous. They were eventually corrected.

The paper (8) on the location of the point of bifurcation along the Jacobian sequence was written during the month following; and some of the early misunderstandings were corrected in the "Note added in proof" to this paper. This paper, together with its companion papers (6) and (7), was completed in June.

Norman came to Yerkes in June: and during the month he stayed we completed the paper (10) on the polytropes as well as the fuller paper (11) on the Beta Canis Majoris stars.

Norman had been interested in the oscillations of the compressible Maclaurin spheroid. It was clear to me that the only new result that one can obtain here pertains to the "radial" pulsations ... never understood why Norman felt the detailed analysis contained in paper (9) as necessary. Anyhow this paper was also written during this same period.

Summer and Fall 1962

The principal problem that loomed large at this time was the third-harmonic oscillations of the Jacobi ellipsoid. This problem required the evaluation of $\delta\omega_{ij;k}$ in general and their appropriate combinations.

I was somewhat disappointed that the eighteen equations could not be reduced further than into two systems of orders 7 and 8, respectively. A novel point in the solution was the elimination of the first-order virial equations by setting the V_i's equal to zero. A further point is that the virial equations should be supplemented by three solenoidal conditions.

During the fall months, the calculations on the third-harmonic oscillations of the Jacobi and the Maclaurin spheroids were going ahead.

In August, I went to Warsaw to attend the conference on general relativity.

Fall 1962 At this time, the definitions of quantities such as v_{ij}, ω_{ij}, and $\omega_{ij;k}$ seemed to me very important. And this early point of view is expressed well in the published version of a public lecture (12) I gave at Berkeley.

In November, I learned of the award of the Royal Medal. And this award helped to elevate my sinking enthusiasm of this period. Indeed on the flight to London, I worked out the theory of the Jeans' spheroids by the virial method.

While in England, I went to Cambridge to visit Lyttleton and Lynden-Bell. But my visit with Lyttleton was to leave a permanent feeling of distaste.

Winter 1962 By January the papers (13, 14) on the third-harmonic oscillations of the Jacobi ellipsoids and the Maclaurin spheroids were complete; also, the paper on the Jeans' spheroids (15). At this stage I had a clearer idea of the role of the equilibrium relations provided by the virial theorem for the purposes of locating neutral points and points of bifurcation. And the detailed paper (16) on the points of bifurcation along the Maclaurin, the Jacobi, and the Jeans sequences clarifies most of the points at issue. But one misunderstanding still remained; and this was corrected in a "Corrigendum" published later.

A somewhat different problem that Norman and I had formulated for ourselves concerns the non-radial oscillations of stars; the problem was to establish the Schwarzschild criterion by a direct normal mode analysis. The virial method could be applied to the solution of this problem and the paper (17) was completed during a brief visit to Harvard in another connection.

Spring 1963 From this point on, the subject was propelled essentially by a feeling of obligation that the subject of the classical ellipsoids required someone to put it in order. And so I undertook the problem of the Roche ellipsoids without any idea that anything really new would develop. While this work was in progress, I went to Yale to give the Silliman Lectures. The invitation had come a year earlier. At that time, I was enthusiastic about the whole subject; and I had agreed to give the lectures on "The Rotation of Astronomical Bodies." The lecture by Whittaker on "Spin in the Universe" was very much on my mind. But by the time April 1963 came around, I had developed considerable coldness towards the subject; and the lectures were to some extent disappointing. I regained some enthusiasm during my last lecture; and Bill Reid, who had come from Providence for the lectures, thought the last was indeed the best. At the end of the last lecture, Brouwer, in proposing a "vote of thanks" presented me with his copy of Volume III of Darwin's Collected Papers — this copy had been given to him by E. W. Brown, who in turn had received it directly from Darwin. I greatly appreciated this generous act of Brouwer.

Also in April, I gave the lecture on "The Case for Astronomy" to the American Philosophical Society. I was not pleased with my lecture; but the written version makes some amends.

It was also at about this time (March) that I had mentioned to Roberts the importance of getting better limits on the ellipticities of slowly rotating masses than had been obtained before. He took up this problem and in his usual fashion raped it so that the problem came out bleeding. And his results did not really go beyond what had been known. I took up the problem and showed what could be done. Roberts's only comment was, "How do you do it; you make me feel ashamed ...". Anyhow this was the origin of paper (18). A major problem in this area still remains.

Silliman
Lectures

To return to the Roche ellipsoids. I was slow in realizing that dynamical instability does not set in at the Roche limit.

In June of this same year, Ledoux visited Yerkes and spent a week. During his visit, I arranged a series of seminars. And in one of them (on June 17) I talked on "The Roche Limit — So Called." In this talk I pointed out that the Roche ellipsoid does not become dynamically unstable at the Roche limit. Ledoux was quite surprised. Apparently he had asked one of his students (Robe) to examine this problem; and seemed disappointed that I had independently gone ahead with it. In this talk I recalled Ledoux's first application of the scalar virial theorem to rotating and non-rotating stars; and how one of his basic papers was written in 1945 in Stanleyville, Congo; and how I had the opportunity to edit the paper for publication.

Summer
1963
During August I gave several lectures on the virial method at the Boulder Summer School. It was also at Boulder that correspondence with Norman clarified the reason why the Roche ellipsoids do not
Roche
become unstable at the Roche limit. His short paper, following mine, was relevant four years later when the problem of the secular stability of Roche ellipsoids was finally solved.

In September, I wrote out my Boulder Lectures. Here I had the first opportunity to clarify to myself the central role of Ferrar's potentials for these developments.

In September, I developed also a general variational principle (21, 22) applicable to rotating masses. While Norman and I undertook the application of this principle to improve our earlier paper (based on the virial method) to the non-radial oscillations of stars, I let Ostriker and Clement make the principal applications. Ostriker applied it successfully to the problem of oscillations of compressible cylinders, while Clement wrote a series of three very good papers on the extension of the principle to rotating stars.

Fall 1963
At this time, my interest in general relativity was increasing. But I felt that the subject of the classical ellipsoids should be completed with a "final" re-examination of Darwin's problem. The formal theory was not difficult: indeed, it is straightforward. But for the first time, I began to appreciate how unimaginative Darwin's work really is: he seems to have had no formal sense nor a compensating physical sense ... I recall Eddington's comparison of Darwin and Poincaré to Pannekock and Unsöld with the implication that Pannekock and Darwin sought to solve a problem by unimaginative numerical computations; and that they did not make any serious attempt to examine the physical (or the mathematical) origin of their results. Eddington used the word "brilliant" with respect to Poincaré and Unsöld. I do not know if I share Eddington's enthusiasm with

respect to Unsöld; but I certainly agree now with his evaluation of Darwin (though at the time I thought that Eddington was needlessly harsh).

Winter
1964

I wrote the paper on the Darwin ellipsoids (23) while on a two-week visit to Stanford (which Schiff had arranged). At this time, I thought that with this paper all of the classical problems had been solved. And the paper (24) which Norman and I wrote for Rosseland's Commemoration Volume was to have been the "final summary."

Indeed with the writing up of my Silliman Lectures in the back of my mind, I turned to the problem of the stability of a rotating liquid drop. I had illustrated this problem in my Silliman Lectures with some beautiful movie films that Dave Fultz had made for me. I found that Rayleigh's work was incomplete. And I knew that the virial method could be extended to this problem. But I was too distracted with many things to concentrate fully on this problem. And the symmetry of the surface energy tensor S_{ij} was proved by Wentzel.

In addition to my papers on the dynamical instability of stars (III, IV, and V) approaching the Schwarzschild limit, I was using my variational principle (papers 21, 22) to complete (together with Norman) our earlier paper (17) on the non-radial oscillations. And during the spring we were working on this paper (25). But an error in my understanding of the variational principle was corrected during the summer. And these corrections together with our move from Williams Bay to Chicago all conspired to make for an exceptionally harassed summer.

An unforeseen event in the spring led to a trail that was to occupy me another three years.

Spring
1964

In the spring of this year, I went to New York to give a talk at the Courant Institute on the virial method. I had arranged to meet Uhlenbeck at the Rockefeller Institute the following morning to discuss with him some ideas on the statistical mechanics of gravitating particles. This discussion did not lead anywhere; but as I had the afternoon free I went to Stechert's book store. And browsing through the new books, I found the Dover reprints of Basset's two volumes on Hydrodynamics. I bought them and glancing through these volumes on the flight back to Chicago, I found that Basset had a chapter on liquid ellipsoids. And here in Basset for the first time I saw references to the work of Dirichlet, Dedekind and Riemann. I immediately realized that there was much more to be done with the virial method. But I laid my thoughts aside hoping to return to these topics at a later time.

Summer 1964

Now during the summer of 1964, when we were making our final move from Williams Bay, I returned to Basset. At no time did I really read Basset's account. I "looked" into his account only to find out what the problems were that Diriehlet, Dedekind and Riemann had formulated. To get into the subject, I first considered the Dedekind ellipsoids. The proof that the geometry of these ellipsoids is the same as that of the Jacobi ellipsoids was simple enough. But I was not prepared to find that the frequencies of the second-harmonic oscillations of the Dedekind and the Jacobi ellipsoids are the same. The point of bifurcation along the Dedekind sequence was clearly of interest to distinguish the two sequences. And the virial method was most suited to the solution of this problem. Indeed, the method shows itself to its best advantage in these latter contexts.

In November of 1964, I spent a week in Harvard where I gave lectures on relativity and on the ellipsoids.

The analysis of the equilibrium and the stability of the Dedekind ellipsoids were completed during the fall (paper 27). The paper on the rotating liquid drop (paper 26) was also completed at about the

Winter 1964

same time. My interests during the fall and winter were primarily in developing a post-Newtonian hydrodynamics. Nevertheless I was propelled into studying the Riemann ellipsoids in the special case when the directions of $\vec{\Omega}$ and $\vec{\zeta}$ were parallel.

1965

1965 was to be a busy year. First there were the two weeks of lecturing at Columbia. Then there was the trip to Newcastle upon Tyne; and finally there were the preparations for the London

Conference on Relativity at which I was to give an invited talk. But the calculations pertaining to the Riemann ellipsoids were continuing. While the paper (28) was completed in April, prior to my departure to Newcastle, and even though I had most of the analysis completed for the case when $\vec{\Omega}$ and $\vec{\zeta}$ are not parallel, I am afraid that the paper as I originally submitted it had several misunderstandings.

At Newcastle I gave a lecture on the historical background of Riemann's problem; and of the particular role that Basset's book had played in introducing me to the subject. Sydney Chapman, who was in the audience, told me later that he was pleased with my references to Basset: apparently Basset had been looked down upon by Lamb and others. But Basset was certainly more sensitive to the epoch making character of the work of Dirichlet, Dedekind and Riemann than Lamb was.

It was only after my return from the London Conference that I explicitly isolated the two bounding self-adjoint sequences. (Norman had discovered these sequences independently while lecturing on these topics in Liége.) But I also found that the S-type ellipsoids become unstable along the self-adjoint sequence $x = -1$ (which he had not). The relation of these results to the equilibrium of the ellipsoids of type III was still not clear. However, all these facts and their inter-relationships became clear subsequently while discussing these matters with Norman. He was most perceptive even though at this time we had effectively ceased the day-to-day collaboration of the earlier years.

1966   The study of the ellipsoids of types I, II and III was carried out intermittently during 1965 and the second paper on the Riemann ellipsoids (paper 29) was completed only in February 1966. But there was still the question concerning the discrepancies between the results of my stability analysis and Riemann's statements in his paper. Norman who had studied Riemann's paper clarified the matter beautifully; and his paper follows mine in the *Astrophysical Journal.*

With the completion of the work on the Riemann ellipsoids the subject had indeed come to an end. And so when I was asked to give the opening address at the inauguration of the Warren Weaver Hall housing the Courant Institute, I chose "Ellipsoidal Figures of Equilibrium" for my subject. The substance of this lecture was later published (30) and effectively forms the first chapter of the book I was to write.

I now began to think seriously of writing up my Silliman Lectures; but devoting it exclusively to ellipsoidal figures. I discussed the usefulness of writing such a book with both Bill Reid and Norman. They thought that I should and that it would be most worthwhile; but what else could they say.

But I was too preoccupied with developing my post-Newtonian work that I laid the matter aside.

1967   In the winter and spring of 1967, I became interested in the variational formulation of the axisymmetric pulsation oscillations of a rotating mass — this interest originated in a parallel interest in the corresponding problem in the post-Newtonian theory. And since Norman was interested in a related problem at the same time, we agreed to work on this subject together.

In early 1967, one further lacuna appeared in the subject of the ellipsoidal figures: I had not discussed the fourth-harmonic oscillations; and these are relevant for the post-Newtonian theory of the Maclaurin and the Jacobi ellipsoids which I had completed by this time. So with considerable resistance, I carried out the long and laborious calculations — Donna was of great assistance in the reductions of the expressions for the various matrix elements — and this paper (31) was finally completed in August. And my paper (32) with Norman on the axisymmetric oscillations of uniformly rotating mass was also completed at about the same time.

The time was at last at hand to write up my Silliman Lectures.

Chapter I was to be an historical introduction. My lecture to the Courant Institute in March 1966, with suitable modifications, became this chapter.

Chapter II on the virial method had to be written and planned with the knowledge acquired over the years; and all of the basic formal developments had to be gathered together here. And this chapter was not easy to write. In contrast, Chapter III was not difficult to organize since my Boulder Lectures provided a basis.

By October, the manuscript for the first three chapters were completed prior to the Michigan 150th year anniversary celebrations. (In preparation for the "celebrations" I had to learn about gravitational collapse; and so had given a number of seminars on this topic in August and September.)

Returning from Michigan, I started on Chapter IV on Dirichlet's problem. I found Norman's Liége lectures on "The Riemann ellipsoids" extremely useful in writing this chapter. In fact, Norman had succeeded in formulating Dirichlet's problem so compactly that I call it the "Riemann–Lebovitz formulation" in this chapter. However, I had to demonstrate the relation to the virial theorem. Chapter IV was completed before leaving for Liege and Rome in November.

At about this time, I had seen the paper by Camm on the virial theorem in stellar dynamics. It seemed to me that Camm had accomplished very little. And so I asked Lee[h] to investigate the matter along the lines which seemed obvious to me. And this paper was also written up before we left for Europe. I sent the paper (33) to the Monthly Notices.

In Liége I learned that Robe had been investigating the secular stability of the Roche ellipsoid by Poincaré's method. The idea of applying the virial method — along Rosenkilde's generalization — which I had laid aside for sometime became an urgent one. And so on returning from Liége, I worked out this theory and discovered to my astonishment that the Roche ellipsoid is secularly unstable precisely between the Roche limit and the point of onset of dynamical

December
1967

---

[h]Edward Lee did his graduate work for the Ph.D. with me during the years 1966–68. When he came to see me for the first time to find out if I would be his sponsor, I suggested that he might be interested to explore the extension of my virial methods to stellar dynamics; and asked him to look through my Boulder Lectures. He came back a week later to say that he found the Boulder Lectures very "boring!".

instability. Aizenman was helpful in securing the necessary integrations in the small interval between the two limits. I wrote this paper (34) for the Ananda Rau Memorial Volume.

Since I was going to La Jolla in January, I wanted to complete Chapter IV on the Maclaurin spheroid before the New Year. With some effort this was accomplished.

Winter 1968

The organization of Chapter V on the Jacobi and Dedekind ellipsoids required that I locate the point of bifurcation by an actual construction of the pear-shaped figure. The concept of linear independence modulo the ellipsoid that I had developed earlier in the post-Newtonian context was most helpful in this construction.

The $(n - 1)$ draft of Chapter V was written in La Jolla.

In February and March and early April, I worked continuously to complete Chapters VI, VII and VIII.

The book was written under protest. But it was written; and by a curious accident, I was able to hand the manuscript personally to the Yale University Press on the fifth anniversary of the last of my Silliman Lectures.

June 1968

Postscript:

Summer 1968

I had supposed that with the handing of the manuscript of my *Ellipsoidal Figures of Equilibrium* to the Yale University Press, I had come to the end of this particular trail. But it was not to be, for I continued to be vaguely irritated by the "sour note" on which the book had ended: the last section of the book on the Darwin ellipsoid concluded with the statement that the matter of the stability of the congruent Darwin ellipsoids was unresolved. And I felt that it was incongruous for a book, claiming to have completed the theory of the classical ellipsoids, to end with a confession that there was yet a further problem to solve. While I was gnawed by this fact, I did not do anything about it during the spring and the summer: I was too preoccupied with the problem of determining the conserved quantities in the second post-Newtonian approximation; and moreover, I had to prepare for the Nehru Memorial Lecture I was to give in New Delhi in November.

    In September (after our return from Seattle where I was lecturing at the Battelle Institute) I began to think seriously about the problem of the stability of the Darwin ellipsoid and how it should be resolved. I had by now decided that a last section on the solution to this problem *must* be added to the book; and there was no time to lose since the proofs of the book were supposed to start coming in by December.

    I first tried to persuade Norman that he should think about this problem stating that we (meaning he and I) had an "obligation" to solve it. He was not persuaded; and added that for his part he felt no particular sense of obligation. So left to myself, I first considered the problem of how one can isolate the configuration which could be quasi-statically deformed without violating any of the equations of equilibrium which determine the Darwin sequence. What was called the Roche limit in my original paper did not isolate this neutral point. Formulated with this limited objective, the problem was not difficult to solve. And the neutral ellipsoids that were isolated were indeed at the distance of closest approach. At the same time, the solution to this problem disclosed the nature of the coupled oscillations that must be considered to settle the question of the stability once and for all. I hastily derived the required characteristic equation before leaving for India so that the calculations could be carried out during my absence. The analysis is one with many pitfalls — more indeed than I realized at the time — and I asked Norman to check it.

    On my return from India, and after completing the work on the second post-Newtonian approximation, I started writing the last section of the book. And while writing out the theory, as I had developed it before I went to India, I discovered that I had made a serious oversight: I had not allowed for the relative motion of the centers of mass of the two ellipsoids. I corrected for this error and it enlarged the order of the characteristic matrix from four to six. By this time, I was so aware of the possibility of errors that I decided to have the analysis scrutinized by Maurice Clement. He promptly discovered that I had made another serious oversight: I had "simply" overlooked the fact that the "coriolis term" resulting from the variation of Ω with t should be evaluated with respect to the common center of mass of the two ellipsoids. The characteristic equation had to be corrected once again. And when the characteristic roots were determined after all these corrections had been made, it emerged — to my total surprise — that the entire Darwin sequence is unstable. I was so suspicious of the result that I had the calculation checked at every stage. (I am afraid that all these frequent corrections tried Donna's patience sorely.) The paper (35) and the last section of the book were thus written; and the "sour note" on which the book had originally ended was thus eliminated.

The question why the congruent Darwin ellipsoids were unstable was at first a puzzle. As I said it was unexpected: Norman Lebovitz, Jerry Ostriker, and Maurice Clement were as surprised as I was. However, it gradually dawned on me that the result, far from surprising, is indeed natural under the circumstances: the instability clearly arises from resonant forced oscillations induced on one by the natural oscillations of the other. Since Donna had already made extensive calculations on the requisite natural modes, it seemed worthwhile to fill in this last lacuna in the theory of classical ellipsoids. The formal theory of these resonant forced oscillations is very simple; but the necessary numerical work is fairly prodigious. The calculations were completed by early summer. But I got around to writing the paper (36) only in October.

I do not believe that I shall be writing any more papers on this subject. And I should be receiving printed copies of *Ellipsoidal Figures of Equilibrium* (37) in a few weeks.

November 1969

GENERAL RELATIVITY (1962–1969)

My first introduction to general relativity was through Eddington's lectures in 1931. And I have still retained the enthusiasm I then acquired for Eddington's mathematical theory of relativity.

In 1933 during one of our walks, Dirac told me that if he were interested in astrophysics, he would want to work on cosmology; and he asked me why I was not interested in the subject. I recall that my reply was "I would rather have my feet on the ground." My principal reason for this reluctance to get seriously interested in relativity was the hardly veiled contempt, I could sense, which physicists like Bohr and others had for the work of Eddington (on fundamental theory) and Milne (on kinematical relativity). And much later, during a visit by McCrea (to Chicago in 195?), Wentzel asked me why I had never worked on relativity. My reply, half jocularly, was that relativity had proved to be the graveyard of many theoretical astronomers and that I was not prepared for burial — not just yet. And, in a more serious vein, I added that astronomers who went into general relativity were prone to play for high stakes; and that my own approach to science was more conservative ... However, when I was writing *Hydrodynamic and Hydromagnetic Stability* I began to think about the field to which I should turn next; and I talked about it to some friends, particularly Gregor. I asked him what he thought of my venture into general relativity. He said, "Why not," and when I expressed my doubts, long entertained, he said, "What can you lose? If your efforts do not succeed, does it really matter? Why not pursue what you wish to." And so gradually I came to the view that I should spend the following years on general relativity: first learning and then exploring if one with my background could make any pertinent contributions.

And so during the summer of 1960 I began my study; and I started earnestly with a series of some twenty seminars on Schrodinger's Space-Time Structure, on Riemannian geometry (Weatherburn), and cosmology (Tolman). And during the fall quarter, I gave my first 400-course on general relativity.

But as it turned out, my intention of 1960 was to be frustrated since Norman's success in solving the problem of the oscillations of Maclaurin spheroid *exactly* was to lead me astray; and the classical ellipsoids were to absorb much of my time during the following eight years. Still, I kept up my study of relativity intermittently, and I continued to give my 400-course on relativity every year. And twice (in 1963–64 — and again in 1967) the course initially scheduled for a quarter was extended to two quarters by petition by the students. And my efforts to contribute to the subject were not entirely in vain.

Jan.
1961

In January 1961, Jim Wright got interested in my account of Gödel's universe in my lectures during the fall quarter. And we worked out together the geodesies in Gödel's universe. This was my first paper (1) in general relativity.

Summer
1962

During 1962 and 1963, I was too occupied with the classical ellipsoids to do much thinking in general relativity. However, largely to get a personal feeling for what the relativists were thinking about, I went to attend the Warsaw Conference on general relativity. The National Science Foundation gave me a travel grant to go to the Conference though in my application I had stated that I was not an expert, that I was not giving any invited talk, that my object in wanting to go to the Conference was merely to get a feeling for what the "experts were thinking," and finally, that if a travel grant was not awarded, I simply would not go!

Spring 1963

By spring of 1963, it became clear that the application of the virial method to the ellipsoidal figures was becoming routine (at that time I thought that only the Roche and the Darwin ellipsoids remained to be treated); and I began to devote more serious attention to general relativity. And I suggested to Contopoulos (who was spending that winter and spring with me) that we might derive the virial theorem appropriate for the post-Newtonian equations of Einstein, Infeld and Hoffmann. The derivation was not too difficult, and we wrote a short paper (II) on it.

Fall 1963

An unexpected opening appeared in the fall of 1963 when I was lecturing on the Schwarzschild interior solution to my class in general relativity. I was aware of the papers Iben and Michel had sent to the *Astrophysical Journal.* Their considerations appeared to me too heuristic; and it occurred to me early in December that it would be quite straightforward to develop the analogue of Eddington's pulsation theory in the exact framework of general relativity and solve a basic problem rigorously. Indeed it *was* easy, but in my "excitement" I made an error in writing the condition for adiabatic change (following the motion) in the framework of general relativity.

Winter 1963

I discussed the matter with Wentzel; but he was not critical enough. And the Letter (III) I sent to the *Physical Review* in January 1964 contained an error. The error became clear to me soon after my return from Stanford in February. I made the necessary correction during a hectic week and sent an erratum to the *Physical Review Letters.* Fortunately, the erratum was published just in time: Misner, Zapolsky and Fowler were already on the trail.

It was at this stage that my closer association with Tooper began.

Tooper, who was interested in general relativity, had seen me during the summer of 1963 with respect to his Letter to the *Physical Review* that had been rejected. Also, he was interested in getting a sponsor for his thesis. Tooper made a very good impression on me. And so when in January 1964, I had found the variational method for ascertaining the stability of general relativistic configurations, I returned to my old problem of the dynamical stability of the white dwarfs approaching the critical mass; and Tooper and I collaborated on a small investigation of this problem (IV). And as one should expect, the configurations become unstable already at moderate densities.

The discovery that dynamical instability sets in already when the relativistic corrections are small suggested that I should obtain the equations of hydrodynamics in the post-Newtonian approximation. With this in view, I included an account of the Einstein–Infeld–Hoffmann theory in my spring lectures on general relativity. (This second quarter of these lectures was given in response to a petition signed by some thirty students.)

The spring and summer of 1964 were distracting with our move to Chicago effected in three stages: from 5550 Dorchester to 4800 Chicago Beach Drive in April, the transfer of the *Astrophysical Journal* to Chicago in July, and finally the complete move from Williams Bay in October. And with all this, I was working on the rotating liquid drop and was starting my investigations on the Dedekmd and the Riemann ellipsoids. And so it was in November of 1964 that I seriously turned to the problem of deriving the

equations of hydrodynamics in the post-Newtonian approximation. I had already worked through the Landau–Lifsehitz treatment of the Einstein–Infeld–Hoffmann theory. But I did not know how their definition of the Lagrangian was to be extended to hydrodynamics. And so I spent most of November reading Fock's book; but I did not find

that very helpful. And suddenly I realized (in early December) that I could follow Landau and Lifschitz in solving for the metric coefficients; and then use the Bianchi identities to obtain the equations of motion. Given this idea, I was able to derive the basic equations during the Christmas holidays.

By the end of January, I had also derived a variational principle for treating the non-radial oscillation of a gaseous mass in the post-Newtonian approximation. And I sent my Letter on this subject to the *Physical Review* on January 25 (VI). During the following months, my principal effort was towards getting my three papers (VII, VIII, and IX) completed before the London Conference on General Relativity. I had been invited earlier to give one of the principal hour talks at this Conference; and I decided now to give it on my post-Newtonian work.

Spring
1965

Spring was a busy time even otherwise. I was completing my first paper on the Riemann ellipsoids. And I had agreed to spend two weeks in Columbia; there was a trip to Newcastle in April pending; and then there was the London Conference.

At Columbia I gave three lectures on the post-Newtonian theory. At the first of these lectures, I was surprised to find van de Hulst in the audience. I told him, "Henk, you have not heard me since the late 40's when you and I were interested in radiative transfer. I am afraid my interests have strayed very far from those common interests of ours. I am afraid you will not recognize me." Later that evening, we were to drive out with Henk to the Spiegel's for dinner. On the way Henk said, "Chandra, contrary to what you said, I should have recognized you at the lecture, even had I been blindfolded and even if your voice had changed to an unrecognizable extent."

At Newcastle in April, I gave similar lectures. I was delighted that Cowling came from Leeds to hear them.

Though under considerable pressure, I did complete my three papers on post-Newtonian equations before leaving for the London Conference and for a week's holiday in Switzerland afterwards. But as a consequence of this pressure, I am afraid that I came to the wrong conclusion with respect to the validity of the Schwarzschild criterion in general relativity. My formulae were correct, but I had misinterpreted them. Bondi seemed to sense that something was wrong.

The London Conference was quite stimulating. I had pleasant exchanges with Synge, Fokker and Fock.

Summer 1965

Returning from Switzerland, I was preoccupied with correcting my errors in the post-Newtonian and the Riemann ellipsoid papers. And then in August, there was the summer school at Cornell where I had to give some lectures.

Fall 1965

In the fall of 1965, I began to think about solving the problem of the Maclaurin spheroids in the post-Newtonian approximation. There were unexpected difficulties. These were concerned with the

Winter 1965

question of the displacements that should be used to deform the Newtonian figure to obtain the post-Newtonian figure. By a series of trials and errors, I finally got on to the essential concept of displacements linearly independent modulo the spheroid. And the problem was finally solved.

Spring 1966

One unexpected result which emerged was the *absence* of a finite solution at a certain definite point along the Maclaurin sequence. I realized that this point must be a point of Newtonian Instability of the Maclaurin spheroid for fourth harmonics. (And this was the reason why I went into a detailed analysis of the fourth-order virial equations.)

Summer 1966

Once the problem of the Maclaurin sequence in the post-Newtonian approximation had been solved and completed (paper X, June 1966) it was logical to proceed with the corresponding theory for the Jacobian sequence. This problem also created some fresh problems; and some unexpected identities had to be established. The analysis, eventually, became a *tour de force*; but I have the feeling that no one really cares. To a large extent, these problems could be solved only by one with the experience I had gained with classical problems. But the methods are far too special and I doubt if anyone has made the effort to understand the results. The paper (XII) on the Jacobian sequences was completed in October.

In August, I was occupied with a problem I had suggested to Contopoulos already in 1963. Is there a transformation to the center of mass in the Einstein–Infeld–Hoffmann theory? Contopoulos had effectively solved the problem; and since our original understanding was to write it jointly, Contopoulos wanted it that way. But when I started working through Contopoulos[1] derivations, I found that a number of obscurities had to be resolved. And their resolution took some time. I finally sent the paper to the Royal Society (XI).

Winter
1966

Originally, it was my intention not to give any lectures during the academic year 1966–67. But I changed my mind and decided to give a course on the mathematical foundations of relativity during the winter quarter.

These lectures were to consist of the elements of topology and differential geometry. And while giving these lectures, I had weekly conferences with Saunders Mac Lane. He was most helpful in initiating me into the spirit of the modern methods. By petition of the students, these lectures were continued during the spring quarter. In these later lectures, I used Trautmanf's Brandeis Lectures as my principal text.

Trautman visited me for a few days in May. And Penrose spent three weeks in June and July. Trautman gave three lectures on his ideas on conservation laws in general relativity; and Penrose gave a series of six lectures on his method of spin coefficients. Their visits were unusually stimulating to me and to my students.

Spring &
Summer
1967

During the spring and summer months I began an investigation of the oscillations of rotating mass in the post-Newtonian theory. But the Newtonian theory had to be understood first. And since Norman was interested in a related problem, we agreed to work together on this phase of the problem. This work was completed in August. Then I worked out the corresponding theory in the post-Newtonian framework. But that paper remains to be written (in June 1968).

Fall 1967

Also during the fall, I started looking seriously into the problem of the post-post-Newtonian theory. The basic equations were derived, together with Nutku, in November; and in December, while in Austin (Texas) for a "Dirac Symposium," I was able to locate the conserved density in the post-post-Newtonian theory.

But the isolation of the other conserved quantities had proved difficult. Moreover, the completion of my *Ellipsoidal Figures of Equilibrium* had been a constant preoccupation during August 1967–April 1968.

Looking back over the past eight years since completing my *Hydrodynamic and Hydromagnetic Stability*, I find that the period has been one of constant distraction, interruption, and frustration. It was not possible to pursue a single theme with a single mind. The alternation between general relativity and the classical ellipsoids, the feeling of obligation interfering with what one wanted to pursue, and the growing responsibility associated with the *Astrophysical Journal*, all had contributed their share.

June 1968

July & Aug. 1968
I had been worried all winter and spring how I was going to determine the conserved quantities in the second post-Newtonian approximation. In the framework of the first post-Newtonian approximation, I had succeeded in isolating them essentially by a process of inspection; this was possible since the equations of motions were sufficiently simple that with some familiarity with them one could feel one's way towards the necessary manipulations. The equations in the second post-Newtonian approximation were far too complicated to hope that one could succeed similarly in *its* context as well. And sometime in May or June, it occurred to me that what was needed was an algorism that will avoid the necessity of intuitive or trick manipulations. Once the need for such an algorism was realized, the idea that one must, for this purpose, turn to the pseudo-tensors was a natural one.

I was first inclined to favor the Einstein pseudo-tensor: I suppose because I had known it! But I soon abandoned it and turned to the Landau–Lifshitz pseudo-tensor: It had the clear advantage that its use required no separate calculation for isolating the conserved angular momentum.

All these facts gradually became clear to me and in July (with my students Nutku and Greenberg safely dispatched to the Brandeis Summer School!) I started on the evaluation of the Landau–Lifshitz complex. Since the idea had first to be tested, it was important to check it in the framework of the first post-Newtonian approximation in which the conserved quantities had already been isolated.

The first quantity to evaluate was of course t^{00} and θ^{00}. On evaluating θ^{00} and comparing it with the conserved matter energy density, $c^2 \rho u^0 \sqrt{-g}$, the need to define the notion of the equality of two quantities modulo divergence became clear. And the fact also emerged that $\theta^{00} - c^2 \rho u^0 \sqrt{-g}$ defines, modulo divergence, the energy conserved in the *lower* approximation: in the first post-Newtonian approximation, the quantity isolated turned out to be the conserved energy in the Newtonian approximation. I then realized that the evaluation of θ^{00} and $c^2 \rho u^0 \sqrt{-g}$ in the framework of the second post-Newtonian approximation will determine the conserved energy in the first post-Newtonian approximation: a circumstance which would provide a valuable check on the whole procedure.

The next quantity I evaluated was $\theta^{0\alpha}$ still in the first post-Newtonian approximation; and it was easy to verify that $\theta^{0\alpha}$, modulo divergence, agreed with the conserved linear momentum π_α that I had isolated earlier.

Then proceeding to the second post-Newtonian approximation, I verified that $\theta^{00} - c^2 \rho u^0 \sqrt{-g}$ evaluated in this approximation did indeed agree, modulo divergence, with the conserved energy of the first post-Newtonian approximation.

All of the foregoing was done during July and prior to going to Seattle. I was therefore able to give an account of these results in my lectures at the Battelle Institute. I thought that Dyson was perhaps the person most responsive to my approach. Kip was gently critical; Bardeen seemed to indicate that he knew it all; and Taub was plainly irritated that he had not thought of it all in the first place: they were, as he said, "almost implicit" in his papers.

Returning to Chicago in late August, I asked Nutku first to check all the calculations I had carried out during his absence. Then we started on the mountain of calculations that had to be scaled to isolate all of the conserved quantities in the second post-Newtonian approximation. The order in which we carried out the calculations was about as follows:

(i) Since θ^{00} had already been evaluated, $\theta^{0\alpha}$ was clearly the quantity next in order of priority. Evaluating it, I was surprised that it could not be reduced, modulo divergence, to a quantity confined only to the volume occupied by the fluid. The question arose: Is this a peculiarity of the particular gauge in which the field equations had been solved? Or might it be that in another gauge $\theta^{0\alpha}$ could be reduced, modulo divergence, to a quantity confined to the volume occupied by the fluid? The dependence of the conserved quantities on the choice of gauge was therefore to be investigated.

It was not difficult to show that $\theta^{00} - c^2 \rho u^0 \sqrt{-g}$ was independent of the gauge chosen in the second post-Newtonian approximation; this independence is clearly necessary since it refers to a quantity conserved in the lower approximation. But on evaluating $\theta^{0\alpha}$ in a general gauge, it seemed at first that by no choice of gauge can the conserved momentum be reduced, modulo divergence, to the volume occupied by the fluid. This was an error that was corrected later. A gauge with the required property does exist. Correspondence with Professor Stachel was helpful in eliminating the original error.

(ii) We next examined how the derived equations of motion depended on the choice of gauge. To our surprise, we found that the equation as derived was formally valid in all gauges.

(iii) In view of the fact that the conserved linear momentum was not, in the chosen gauge, confined to the fluid, the question of the relationship between the equations, $\theta^{\alpha i}_{,i=0}$ and $T^{\alpha i}_{;i=0}$, arose. Again the relationship was first studied in the framework of the first post-Newtonian approximation since $t^{\alpha\beta}$ appropriate to this approximation required a knowledge of the Christoffel symbols in the second post-Newtonian approximation.

The (α, β)-component of the pseudo-tensors appropriate to the first post-Newtonian approximation, i.e. to $O(c^{-2})$ was therefore evaluated using the full knowledge of the Christoffel symbols determined in the second post-Newtonian approximation. And the remarkable fact emerged that all terms of $O(c^{-2})$ in $t^{\alpha\beta}$ which were derived from an explicit knowledge of the second post-Newtonian approximation cancelled separately. This was clearly a result of considerable significance for the evaluation of $t^{\alpha\beta}$ in the higher approximations. A further fact that emerged was that in the first post-Newtonian approximation the equations satisfied by $\theta^{\alpha i}$ and $T^{\alpha i}$ were identical.

(iv) The next question that was to be decided was whether the identity of the equations satisfied by $\theta^{\alpha i}$ and $T^{\alpha i}$ was maintained in the second post-Newtonian approximation as well. The evaluation of $t^{\alpha\beta}$ to $O(c^{-4})$ is very long and tedious; and the reduction of the equation $\theta^{\alpha i}_{,i=0}$; inclusive of terms of $O(c^{-4})$) equally long. They were carried out but the accidental omission of a single term in the evaluation of $\theta^{\alpha\beta}$ resulted in the equations satisfied by $\theta^{\alpha i}$ and $T^{\alpha i}$ being different. The possibility that this discrepancy may be due to an error in the evaluation of $\theta^{\alpha\beta}$ was first suggested to us by Dr. Estabrook. When the error was corrected the equations did agree.

(v) The last remaining quantity to evaluate was the conserved energy in the second post-Newtonian approximation. The evaluation of this quantity requires the knowledge of $g_{\alpha\beta}$ appropriate for the third post-Newtonian approximation. Most of the requisite calculations were carried out prior to our departure to India. What was left was completed after our return. And the entire theory was ready to be written up by Christmas.

I should add that during the fall quarter I had devoted a larger part of my course on relativity to post-Newtonian methods.

Jan.
1969

Before I could start writing up my papers on the conservation laws in general relativity and on the second post-Newtonian approximation, I had to write first the paper on the Darwin ellipsoid and the last section of the book. The discovery at this time of errors in my analysis of the coupled modes of oscillation of the Darwin ellipsoid was a source of considerable distraction. Also my Richtmyer Lecture had to be written up before the end of January. With all these pressures, only the paper on the conservation laws (XV) was written in January.

Feb.
1969
The organization of the paper on the second post-Newtonian approximation was not an easy matter: some 400 odd pages of calculations had to be abstracted and summarized. Most of February was taken up with it.

My paper on the second post-Newtonian approximation (XVI) was written jointly with Nutku. I ought to state here that Nutku's collaboration with me on this paper was effective and essential. Nutku was enthusiastic and keen; and much of the long and laborious calculations were carried out and checked by each of us in turn. With the many distractions I had to contend with, and with my low spirits generally, I doubt if this long and difficult work would have been completed without Nutku's youthful enthusiasm.

March
1969
In March, I wrote out the variational principle I had derived (a year earlier) for the oscillations of a uniformly rotating mass in the first post-Newtonian approximation. This paper (XVII) was my contribution to the Wentzel "Festschrift" which Nambu and Freund were preparing.

About this time, all the galley proofs of my *Ellipsoidal Figures of Equilibrium* had also been read; and so it was with some relief that I welcomed my visit to Pisa where I was to participate in a small conference on pulsars: it gave me one free day in Florence.

May &
June
1969
The completion of the second post-Newtonian approximation left the determination of the radiation-reaction terms that must appear in the next one-half-approximation as the last major problem to solve.

Some months earlier Kip Thorne had visited Chicago and had given a seminar on the weak field limit of his exact theory of non-radial oscillations of neutron stars. It became clear from his discussion that the Sommerfled radiation condition at infinity must somehow be applied to start the $2\frac{1}{2}$-post-Newtonian approximation. But I was not in complete sympathy with Kip's approach. I tried to read Peres' papers in which he claimed to have determined the radiation-reaction terms in the framework of the original Einstein–lnfeld–Hoffmann theory in agreement with the predictions of the linearized theory of gravitational radiation. I could make no sense of his papers. At last, I came upon Trautman's 1958 paper in the Bulletin of the Polish Academy of Sciences; and I also recalled his King's College Lecture notes in which the same ideas were also described. Trautman's program made sense to me; and I was convinced about its essential soundness. I could broadly see the outlines of the theory that would emerge. I was extremely anxious to work out some of the details before the Cincinnati Conference on general relativity at which I was scheduled to give an account of my post-Newtonian work.

My confidence in the eventual outcome of the theory was so great that I gave a brief description of my ideas at the farewell seminar organized by Nambu for Wentzel. I had talked about this particular problem to Wentzel over the years; and I thought that it would be appropriate to present the solution at the last of his weekly seminars.

By the time of the Cincinnati Conference I had worked out some of the analytical results; and I gave an outline of the new theory in my lecture. But as it turned out I was at that time under a misapprehension about the total correctness of Trautman's procedure.

July & Aug. 1969

On returning from the Cincinnati Conference, I began to develop the theory of the $2\frac{1}{2}$-post-Newtonian approximation with care. It soon became clear that I was not getting the required agreement with the linearized theory. And I realized that, contrary to my earlier belief, Trautman's final result was also in contradiction with the linearized theory. Clearly there was something wrong. I was puzzled by this discrepancy since I had acquired by this time total conviction in the essential correctness of Trautman's basic ideas. Then on a Sunday (August 24) when I was trying to go over in my mind the different steps in Trautman's reasoning, it flashed on me that he was wrong in working with the energy momentum tensor T^{ij}: he should have been working with the Landau complex θ^{ij}. I recast my results in terms of θ^{ij}; the first result that emerged was most surprising: the $2\frac{1}{2}$-post-Newtonian terms do not contribute to θ_3^{00} and $\theta_4^{0\alpha}$. At first, this seemed to be a catastrophe; but soon it became clear that the vanishing of θ_3^{00} and $\theta_4^{0\alpha}$ are absolutely essential for the logical consistency of the theory!

With the radiation-reaction terms uniquely and unambiguously determined, it did not take very long to show that the predicted rates of dissipation of energy and of angular momentum were consistent with the linearized theory of gravitational radiation.

The last remaining problem was to determine the contribution of the $2\frac{1}{2}$-post-Newtonian approximation to the "conserved energy." Some delicate questions concerning convergence had to be answered. But with the basic confidence in the correctness of the theory, it was not difficult to resolve all of them.

By early September, it appeared that there were no loose threads hanging. And it was possible to write up the final paper (XVIII) on the $2\frac{1}{2}$-post-Newtonian approximation as well as my paper (XIX) for the *Proceedings of the Cincinnati Conference* just in time to allow a week's holiday in Cape Cod before the beginning of the new quarter.

The completion of this paper on the $2\frac{1}{2}$-post-Newtonian approximation brings to an end a project I had formulated for myself in 1962 at the Warsaw Conference on general relativity. And over the years I never lost sight of my basic objective. I only wish that I had not been so constantly distracted that I could have arrived at this stage of my understanding of general relativity a few years earlier.

November 1969

Postscript

Once the radiation-reaction terms had been determined, it was clear that the highest priority had to be given to the solution of the following two problems. (1) How does the Jacobi ellipsoid evolve by virtue of the fact that it must radiate gravitationally? (2) Does the dissipation of energy by gravitational radiation induce, in the manner of viscosity, a secular instability of the Maclaurin spheroid at the point of bifurcation with the Jacobian sequence? (Ostriker had raised the latter question with me some months earlier.) I asked Esposito to look into these problems. His familiarity with the virial techniques was not sufficient to make any progress; and I myself could not return to their consideration before the Christmas recess. And I was anxious to complete the solution of the two problems before the annual American Physical Society meeting in Chicago late in January. (At this meeting I was to chair a session on gravitational radiation.)

Starting to think about the problem during the Christmas holidays, I soon realized that the first problem to resolve was the transformation to a uniformly rotating frame the quantity $d^n \vec{I}/dt^n$ ($n = 2, 3, 4, 5$) occurring in the radiation-reaction terms and expressed in an inertial frame. The method to be used was clearly an adaptation of that (due to Lebovitz) described in Chapter IV of my E.F.E. And I was surprised that in the context of what I needed the required transformation could be written out explicitly.

With the single "technical problem" resolved, there was no difficulty in writing out the equation which determines the evolution of the Jacobi ellipsoid by gravitational radiation. And Donna integrated the equation in two different ways to ensure the correctness. However, the determination of the precise asymptotic behavior as the Jacobi ellipsoid approached the non-radiating state at the point of bifurcation was quite unexpectedly troublesome. The difficulties were eventually resolved after a fruitful discussion over lunch with Norman Lebovitz.

Turning next to the problem of the toroidal oscillations of the Maclaurin spheroid allowing for radiation reaction, I found that the solution could be effected without any real difficulty. But I was concentrating so exclusively on the mode that becomes neutral at the point of bifurcation that after having solved the problem, I overlooked the crucial fact that the mode which becomes secularly unstable is not the one which becomes neutral but the one which acquires the frequency 2Ω at the point of bifurcation. And the first version of the paper I sent to the *Physical Review Letters* stated wrongly that gravitational radiation does not induce secular instability. (A fact of some interest in this connection is that both Jim Bardeen and Kip Thorne thought that gravitational radiation should not be expected to induce instabilities; and they were smug about the result as I then stated.) However, two weeks later I realized my horrible oversight; and a completely corrected manuscript was sent to the Editors. Since the Letter had not been set in type, I thought that the printed version would be entirely correct. But in spite of all the care I took, the proofreader failed to notice that the last sentence of the abstract had been changed. The result was that the abstract in the published version states a result contrary to what is established in the main text. An erratum was published two weeks later. But this unfortunate episode succeeded in killing all the joy that there might have been in establishing a result which I believe is of considerable importance for the problem of gravitational collapse.

April 1970

The Fallow Period (1970–1974)

1970   The demonstration of the secular instability of the Maclaurin spheroid by a Dedekind mode induced by gravitational radiation-reaction, though it was to play an enduring role in subsequent developments, ended anticlimactically (as I have written earlier). It foreshadowed a long period of disappointments of many kinds: but these were yet to unfold.

July
August

With the work on the post-Newtonian approximation carried out as far as I cared, it was necessary to think about the directions in which I should proceed. Partially on this account I arranged a "summer school" with Carter, Ellis and Geroch for the months of July and August.

The two months of the summer school turned out to be very strenuous: with three lectures and a seminar every week. I worked out a full set of notes for the lectures by Geroch and Ellis with the hope that they might be helpful in formulating my own future scientific plans. It did not turn out that way. In fact, already in July it occurred to me that a systematic investigation of axisymmetric perturbations of rotating systems would be a worthwhile undertaking. In particular, I felt certain that a formula for σ^2 for slow rotation, can be obtained, exactly as in the Newtonian theory, i.e.

$$\sigma^2 = \sigma_0^2 + \Omega^2 \sigma_1^2$$

where σ_0 is the value appropriate for the radial oscillations of the non-rotating star and σ_1 depends only on the proper solution ξ belonging to σ_0 and the spherically symmetric part of the distortion caused by the rotation. I was aware that the developments would require elaborate algebraic work and I thought that John Friedman, who was completing his first year as a graduate student, might be helpful. So I explained the program to him asking him if he would like to collaborate with me on the work and warning him that it would be a full two-year project and would require that he accompany me to Oxford in 1972. I also told him that I could not promise to start on the work before October and that he should prepare himself meantime. Friedman was responsive to the suggestion.

To return to the summer school, I arranged a week's recess during the last week of July in order that I could go to St. Andrews to give the opening theoretical talk on white dwarfs at an I.A.U. Symposium arranged by Luyten. (I really did not want to participate; but I felt obliged on account of Luyten.)

September   The summer school ended on the last Friday of August with a luxurious "farewell" tea arranged by Persides. And on Sunday we were to leave for England: to give the opening address to an International Conference on Radiative Transfer at Oxford on the following Tuesday and for a two-week holiday in Cornwall and the Moors. It was a difficult matter turning my thoughts away from relativity and to radiative transfer — a subject on which I had not talked for some twenty years. Finally on Sunday afternoon, I called Norman to come to hear me attempt a rehearsal of my talk. In that way, I thought I could be forced into thinking about the subject or at least get into a more responsive state of mind. Norman was agreeable; and the effort to talk consecutively for an hour was very helpful. That was my only preparation. We left later that evening for London and arrived in Oxford late on Monday afternoon; and my talk was scheduled for the following morning. To my considerable surprise, I found that once I started to talk, the enthusiasm of the forties came back to me. It was probably one of the more "eloquent" lectures that I have given. I concluded by thanking the audience "for their patience with an ancient mariner!"

We left before the end of the conference on our two-week holiday in Cornwall, Dartmoor and Exmoor. The walk along the coast from Larmona Cove to Land's End is still fresh in my memory.

October
November

We returned to Chicago late in September and I plunged into the various matters that had to be attended preparatory to my retirement from the Editorship of the *Astrophysical Journal* by March 31, 1971. I shall not go into these matters here: they form a separate story. At the same time, to prepare myself for the work on axisymmetric systems, I devoted my fall lectures on general relativity to Cartan's calculus. I acquired sufficient familiarity with the modern methods by deriving the Schwarzschild and Oppenheimer-Snyder solutions by the calculus of exterior differential forms. I also worked out the various equations appropriate to stationary axisymmetric systems.

December

By early December I was ready to embark on my projected work on axisymmetric systems. I worked out the various field equations assuming the same form for the metric as in the stationary case but letting the various metric coefficients to be functions of time as well. In other words, I worked with the metric

$$ds^2 = -e^{2\nu}(dt)^2 + e^{2\psi}(d\phi - \omega dt)^2 + e^{2\mu_2}(dx^2)^2 + e^{2\mu_3}(dx^3)^2 ,$$

where ν, ψ, ω, μ_2 and μ_3 were allowed to be functions of x^2, x^3 and t. At this stage I did not realize that this was not the most general form of the metric that one must use. Nevertheless, working with this special case did help in the eventual unraveling of the problem.

January
1971

With the beginning of the New Year, I was anxious to make some progress before I turned to other matters that required attention. So setting $\mu_2 = \mu_3$ Friedman and I tried to derive a variational principle for axisymmetric oscillations of a uniformly rotating star. After many false starts, we were able to cast the expression for σ^2 in a symmetric form. While the analysis was wrong in details —

because we were not working with the most general form of the metric — we did understand the different roles of the initial-value and the dynamical equations. In particular, we understood the roles of the equations ensuring the conservations of entropy, baryon number and angular momentum, besides the others which derive from the field equations themselves. With the variational principal derived — albeit in its restricted form — I felt confident that the back of the problem had been broken. So I asked Friedman to work out the general case (i.e. $\mu_2 \neq \mu_3$) and derive the "general" variational principle. He succeeded in deriving a symmetric form, but it was clear that he was not sensitive to "elegance": he had paid little attention to the requirement that the formulae must at all stages manifestly reduce to the earlier work if μ_2 was set equal to μ_3; and his formulae did not manifestly satisfy this requirement. But I had to put this matter aside and turn toothers which required immediate attention. Besides winding up the affairs of the journal, I had three papers to write before the end of March when we were due to go to India. Two of these papers were related to my earlier work on the post-Newtonian deformations of the Maclaurin and the Jacobi ellipsoids. In the earlier papers, I had not "normalized" the post-Newtonian configurations with respect to the Newtonian configurations for equal angular momentum and baryon number. Towards this end Bardeen had redone my Maclaurin work, *ab initio*, by a new method. I was anxious to show that the renormalization could equally be effected by supplementing my earlier work with some elementary calculations; and also that the same methods could be extended to renormalize the Jacobian figures as well, particularly as Bardeen felt that he was "not up to it." Actually, his methods would not succeed for the Jacobian ellipsoids: they were too special. The requisite calculations had been carried out by Donna during the summer and the fall; and I wished the papers to be written up before the end of March. Besides, a paper for the "Synge-Festschrift" was due at the same time. I had decided during the summer that I would redo my 1938 paper on composite configurations consisting of isothermal cores and

February
March

homogeneous envelopes in the framework of general relativity and show that the "spiraling" of the mass-radius curves one finds in general relativity has a classical basis. The requisite integrations had also been carried out during January and February; and the paper had to be written. These three papers took most of February and March. (I sent a preprint of the paper on the relativistic isothermal gas sphere to Martin Schwarzschild with the inscription, "Dedicated to the memory of Karl Schwarzschild, Robert Emden and Edward Arthur Milne with respectful admiration by the **u** and **v** functions.")

With the three papers written and with the *Astrophysical Journal* changing editorships on March 31, we left for Bombay with April a considerable sense of relief. Apart from a week in Bombay, I May visited Bangalore where Lady Raman showed me around the Ra- June man Institute and Ramaseshan told me in detail of Raman's last days: extremely instructive and touching in many ways.

The two months after our return from India were full of distractions: a course on Mathematical Physics, the meeting of the National Academy of Sciences at the end of April, and the Convocation Address in June. So it was only after the end of the quarter that I could resume work on axisymmetric systems; and earnestly only during the three weeks in July at the Massachusetts Institute of Technology with C. C. Lin's group.

July The first thing that was on my agenda when I got to the Massachusetts Institute of Technology was to reduce the expression for σ^2 into a self-adjoint form when $\mu_2 \neq \mu_3$ — but still not on the most general form for the metric. By carefully following the treatment in the case $\mu_2 = \mu_3$, it was possible to cast the expression for σ^2 in a form which was not more complicated than in the case of equality. This was about all I could do during the three weeks at the Massachusetts Institute of Technology.

Returning to Chicago by the end of July, I had just about a month before leaving for our three months at the California Institute of Technology. And again during this month I had to devote my attention to other matters.

August

First, there was the matter of collating some side work on the application of the tensor-virial theorem to stellar dynamics which I had been pursuing since our return from Cornwall. I had always been interested in the application of the tensor-virial theorem to the problem of collapse of spherical and spheroidal systems. The differential equations for a "cluster" collapsing with a uniform density is easy enough to write. One has to integrate a pair of coupled ordinary differential equations. Donna had been integrating these equations for various initial conditions in her spare time all during the fall, winter and spring. It remained to assemble the results and write the paper; and this I did.

Another paper I wrote during this month expresses my own attitude to the "Derivation of Einstein's Equations." Over the years — i.e. since 1962 — I had gradually developed a personal approach to the matter; and I thought that it might be useful to put it on record. Trautman had encouraged me to write it: he appeared sympathetic to my point of view.

During my time at the Massachusetts Institute of Technology, I had worked carefully through Hartle's paper on the equations governing slowly rotating masses in general relativity. I had done it primarily as a preparation to finding the effect of slow rotation on the onset of dynamical stability. While studying Hartle's paper, it occurred to me that it would be worthwhile to work out the structure of slowly rotating homogeneous masses in general relativity since that would enable one to study the effects of rotation at more intense gravitational fields than are possible with more conventional neutron-star models: homogeneous configurations can exist down to $R = (1.125)2\,GM/c^2$ while neutron-star models become unstable already when $R = (2 - 2.5)2\,GM/c^2$. I had also derived the basic equations appropriate for the case of uniform energy-density at this time. This was to be one of my side projects for the next two years.

September
October
November

By the first of September we were in Pasadena. At long last, I could concentrate on the work on axisymmetric systems which had been interrupted so often. The first thing to do was to specialize the results we had obtained for the case of slow rotation. Several oversights and misunderstandings had to be corrected: for example, the surface integrals which had been so cavalierly ignored had to be included and carefully examined. Also, the result had to manifestly agree with the variational expression for radial oscillations when Ω^2 is set equal to zero. (Even so, the expression was not; reduced at that time to its most practical form in which the integrations are confined to the volume occupied by the fluid; this was accomplished much later.) The next question that had to be settled, once and for all, was the one that had been gnawing me from the beginning, namely, was the metric that I was using the most general one? Actually, some discussions with Trautman earlier in spring had in fact provided me with the right clues; but I did not follow them at the time. When I began seriously to contend with this question, the relevance of Trautman's remarks became clear and I realized at last that the form of the metric with which I should work was

$$ds^2 = -e^{2\nu}(dt^2) + e^{2\psi}(d\phi - \omega dt - q_2 dx^2 - q_3 dx^3)^2$$
$$+ e^{2\mu_3}(dx^2)^2 + e^{2\mu_3}(dx^3)^2 ,$$

with the additional restriction that w, q_2, and q_3 occur only in the combinations

$$\frac{\partial q_2}{\partial t} - \frac{\partial \omega}{\partial x^2}, \quad \frac{\partial q_3}{\partial t} - \frac{\partial \omega}{\partial x^3} \quad \text{and} \quad \frac{\partial q_3}{\partial x^3} - \frac{\partial q_3}{\partial x^2}.$$

The generalization of the earlier work to allow for the additional terms in the metric was relatively straightforward though many of the subtleties became clear only as the months went by.

The next matter to be concerned with was the isolation of the variable which carried the information on the emitted radiation. For this purpose, I evaluated the Landau complex following Cornish's prescription. However, when the time came to return to Chicago, all the questions had not been fully clarified.

The stay at the California Institute of Technology was profitable. I had no interruptions; and the fact that all the astronomers avoided me actually contributed to my recovering some sanity. I did get to know Kip Thome's students, particularly Press and Teukolsky. One unpleasant interlude was my having to go into surgery for a fourth hernia operation.

We returned to Chicago by the first of December. During this month, I tried to write up the work that had been completed up to that time as two papers. The first paper was to assemble the basic equations; and the second paper was to be devoted to the variational principle and its reduction in the case of slow rotation. The second paper was also to obtain the general criterion for the occurrence of a neutral mode. I thought that the latter criterion could be obtained by reimposing the gauge condition $\delta(\mu_2 - \mu_3) = 0$ allowed under stationary conditions. But this was an error that was corrected only later in Oxford. In any event, the two papers were written and sent to the *Astrophysical Journal*, just prior to our departure on 2nd January for our six months in Oxford.

I felt very tired during my first month in Oxford. But the whole quietness of the Oxford atmosphere and the walks down the Thames had an enormously soothing effect.

By February, it became clear that the manner in which we had made the passage to the neutral mode was not satisfactory — in fact wrong! John raised many objections; but it was not clear to me what was precisely wrong. During a visit to Cambridge, we had a long discussion with Carter and Schutz but without resolving the matter. Finally, one day I argued the matter closely with John; and in this way I was able to locate the precise place where the argument had gone astray. Once the error was located, it was not difficult to resolve the entire matter. The solution as given in the published paper was obtained later that same day. The second paper was revised and resubmitted in March.

Before going on to the problem of how to incorporate radiation, it seemed that our condition for a neutral mode specialized to vacuum metrics should yield Carter's theorem. It was John's insistence that there was an identity here that must be isolated that led to the eventual solution. This paper was written and submitted in April. It was also in April that I gave my talk on the stability of relativistic systems at the Royal Astronomical Society.

May
June

Next, I began to think of the Halley Lecture that I was to give in May. The Lecture was to be, "On the Increasing Role of General Relativity in Astronomy." We went to Ireland the week before the Lecture to participate in a function arranged by the Irish Academy in honor of Synge's 70th birthday; and for a short holiday afterward driving across Ireland.

We had left our car at Heathrow; and coming out of the garage onto the main highway, I had a very bad accident: a car coming down the highway crashed into ours. By the sheerest accident, none was hurt. The back of our car was completely destroyed. This was on Saturday evening; and my Halley Lecture was scheduled for Tuesday. It was my intention (prior to leaving for Ireland) to devote the entire Sunday and Monday to writing out the Lecture. Thinking back, I am still astonished, that after the shock of the accident, I was able to concentrate and write my Lecture as I had planned.

During May and June, there were two problems which occupied me. The first was to incorporate properly, in the variational principle we had derived, the surface integrals representing the emission of gravitational radiation. The principal problem here was to obtain the correct asymptotic behaviors of the various metric coefficients; and John and I had resolved most of the difficulties; but there were still some questions relating to the Landau complex. Meantime, it had occurred to me that the condition for the occurrence of neutral modes appropriate for bifurcation by a Dedekind mode could be obtained by a straightforward "transcription" of the analysis of the time-dependent axisymmetric case. There is a simple reciprocity between the equations governing azimuth-independent stationary systems and time-dependent axisymmetric systems. This reciprocity is fundamental. That is about where I had got by the time our Oxford period came to an end.

July
August

We returned from Oxford during the last week of June. Since we were to go to Athens (to give an invited talk at the First

European Meeting of the I.A.U.), Thessaloniki (to visit Contopoulos), and Trieste (to attend the Symposium in honor of Dirac's 70th birthday), I had just about two months to clear up a number of things. The first item on the agenda was to complete Paper IV of the series on axisymmetric systems. This paper was to discuss the way the variational principle is to be used when allowance is made for the emission of gravitational radiation. The basic calculations had already been made; but consistency with the requirements of the Landau–Lifshitz complex had not been established. After some hard thinking, the matter was resolved and it was possible to send the paper off just prior to our departure to Europe.

But the main problem that interested me at this time was to ascertain whether along the Kerr sequence there was a point at which a neutral non-axisymmetric mode of deformation was possible. I had established the condition for this in July and had found that the integrands diverged on the stationary limit. And it was my idea at this time that one must simply set the perturbations to be identically equal to zero inside the ergosphere. I was not sure whether setting everything equal to zero inside the ergosphere by fact was justified; and I wanted to take the occasion of the meeting in Trieste to discuss this question with the relativists. For the same reason, I decided to visit Ehlers in Munich during the week between the visit to Greece and the meeting in Trieste. But the discussions proved inconclusive; and indeed, retrospectively, I believe that I made a serious error in trying to get the advice of others. (I still believe that my original idea is correct and that it must be pursued.) Meantime, John wished to dissociate himself from these efforts since he was not in sympathy with my views — in fact, his part in our collaboration had effectively ceased by this time.

September   Both in Munich and in Trieste, I repeatedly questioned a number of relativists (including Wheeler, Deser, Ehlers, Trautman, Rees and Persides) whether there was anything basically wrong in setting the perturbations identically equal to zero inside the ergosphere and allowing a discontinuity in the second derivatives of perturbed metric coefficients on the stationary limit. The response was mostly to the effect that it was permissible; but it was clear to me that none of those whom I questioned really understood the nature of my concern.

In Trieste, I was to give a talk on the "Astrophysicist's View of the Universe." The talk was essentially a repetition of my Halley Lecture. I also had to give a brief talk on Fermi. From the reactions that were expressed, the talk on Fermi was the greater success!

October
November
December

Returning to Chicago, I had several problems in mind. The first was to complete the paper with Friedman on the "Criterion for the Occurrence of a Dedekind-like Point of Bifurcation Along a Sequence of Axisymmetric Systems." At this stage, it occurred to me that one should develop an analogous criterion in the Newtonian framework, particularly, for differentially rotating systems. I discussed the problem with Norman suggesting that we might collaborate on deriving the criterion. And, of course, I was constantly and continually troubled by my ideas on the Kerr metric. At this time, I also decided to embark on determining the post-Newtonian deformation of the Dedekind ellipsoid.

The year ended with the paper with Friedman completed in its essentials. And Norman had shown that the conservation of u_0 per baryon in general relativity is replaced by the Bernoulli integral in the Newtonian theory; and the prospect of getting a Newtonian analogue of the relativistic criterion seemed assured. On the Kerr metric, I had derived the series expansions on the stationary limit which made the perturbations and their derivatives vanish on the stationary limit. That was how things stood when the year ended.

1973

With the New Year, I decided that I should give priority to completing the two papers (one with John and the other with Norman) on the criterion for a Dedekind-like point of bifurcation on the relativistic and the Newtonian frameworks. And there was also pending the review of the Wigner–Salam book (dedicated to Dirac) for *Contemporary Physics*. None of these tasks elicited any enthusiasm. With reluctance, I was able to complete these three "assignments"

April

during the month following the end of the quarter.

Besides, during the winter I had thought of generalizing the work with Norman to obtain a criterion for the onset of dynamical instability of differentially rotating stars by a quasi-stationary analysis. I asked Norman if he would care to continue to collaborate with me on this extension. But he was not interested. Also the post-Newtonian analysis of the Dedekind ellipsoids was not going too well: the analysis turned out to be far more subtle than I had thought; and my progress was through making one error after another. And on the Kerr metric, because I was not finding much support for setting the perturbations inside the ergosphere to be zero, I thought I would continue the solutions inside and see if the boundary conditions on the horizon could be satisfied — and this attempt, as it turned out, was a mistake.

May   In May, I had to go to Delhi in connection with the 100th anniversary celebrations of the University of Delhi. Returning from Delhi, I wanted to clear my desk before getting into the stability of the Kerr metric. First, I concentrated on deriving a variational expression in terms of which one can isolate the onset of dynamical instability along a sequence of differentially rotating stars. The required extension of the earlier results with Norman turned out to be very slight. On this account, I thought the paper should be a joint one; but Norman declined. After writing this paper, I turned to the

June
July
problem of slowly rotating homogeneous masses in general relativity. I had started on this problem two years earlier. While at Oxford, I had suggested to one of Sciama's graduate students, Miller, that he could get his teeth into relativistic astrophysics by working with me on this problem. Since I had already worked out the formal theory, his problem was to understand the theory as I had worked out, make it more explicit, and put the equations on a computer. Miller was very conscientious and analyzed the problem, both analytically and numerically, much more thoroughly than was my intention. In any event, he had done more than his share and by June he had sent me all of his work. I collated the entire material and sent the paper to the *Monthly Notices*. By the time the two papers were written, we

had to leave for Aspen where we were to spend a month during July and August.

August
In Aspen, I concentrated on reducing the variational principle we had derived appropriately for the non-radial oscillations of stars and for the Schwarzschild black hole; and looking further into the stability of the Kerr metric. I wanted to get my ideas clear in preparation for the talk I was to give at the Copernicus Symposium in Warsaw in September. I made some progress.

The peace I had hoped for in Aspen was shattered by Lalitha's accident resulting in a broken ankle. It happened while we were up in the mountains; and with no one around to help us, it was a shattering experience. Fortunately, the break was a clean one and the accident was not by any means as serious as it might have been. But the accident did require our cancelling the holiday in Poland that we had planned with the Trautmans, after the Symposium in Warsaw.

After returning from Aspen, I had time only to write out my Warsaw talk; and the matter of the Kerr metric was left inconclusive. It was clear from the reaction to my talk at Warsaw that I was alone in thinking that the Kerr sequence may bifurcate at some point.

October
Returning from Warsaw, I finally decided to write my article on black holes for *Contemporary Physics* which had been promised for the preceding Christmas.

I did not feel sufficiently enthusiastic to go "it my way" in the Kerr problem; and as a result a long period of inactivity followed. The only positive thing I did do during the fall was to finally resolve the matter of the post-Newtonian Dedekind ellipsoids. The year ended, as it began, with doubts and frustrations. And I found myself in the hospital by New Year's Eve with angina pectoris.

1974
The New Year began with the realization that my attempt to conform with the "establishment" in continuing the solutions, which
January
vanish on the stationary limit, into the ergosphere was a mistake: John had meantime shown — to his satisfaction and apparent delight — that the non-radiative solutions I had found inside the ergosphere can be eliminated by a gauge transformation. Clearly, I should have

stayed with the position I had taken prior to the Trieste meeting. Now it looked lame to go back to that position; but that is really what I should have done and what I needed to do. But my "spirit" had been broken; and I did not feel like embarking on the large scale computing that would be necessary to settle the question. So that was that.

February
March

In February, I was able to write up the paper on the post-Newtonian Dedekind ellipsoids. It was a long haul; and the final result that the solution diverges very soon after the sequence bifurcates from the Maclaurin sequence was a surprise; and it could prove to be an important result to have established.

With the coming of spring, I wanted very badly to shake off the mood of pessimism that had gradually engulfed me during the preceding eighteen months. It seemed best to discontinue my efforts towards proving the instability of the Kerr metric. I had a feeling of being imprisoned in my own ideas; and I wanted to break away. Finally, one day in March, it occurred to me that I should try to identify the Zerilli function in my gauge and relate it to the Bardeen–Press equation, i.e. the equation to which the Teukolsky equation reduces in the limit $a = 0$.

April
May

At first I tried the Regge–Wheeler method of finding a linear relation among the defining scalars by requiring that the second-order field equations follow from the first-order equations. I found that this led nowhere: the second-order equations were identically satisfied by virtue of the first-order equations. Finally, I used the gauge transformation relating my scalars with the Regge–Wheeler scalars to find out what Z is in terms of mine. Once the identification was made, it was straightforward enough to check that it satisfied the Zerilli equation. The next problem was to relate the Newman–Penrose $\delta\Psi_0$ with Z. Fortunately, while at Oxford, I had checked John's calculation relating $\delta\Psi_0$ with the scalars in our gauge. After some very elaborate calculations, it was possible to relate $\delta\Psi_0$, rather a function Y (which is $\delta\Psi_0$ times a simple factor), with Z and its derivative. The problem was now to derive the Bardeen–Press equation from the Zerilli equation. It was at this stage that some extraordinary identities which lie concealed in the theory were disclosed.

June   Meantime, I had also been thinking how to determine the quasi-normal modes of the Schwarzschild black hole. Steve Detweiler had independently been working on this problem by directly integrating the Zerilli equation. It seemed to me that a better way would be to transform the wave equation to the Riccati form. Detweiler agreed to collaborate with me on this problem; and during the month of July he was able to complete this work. We agreed that after his summer holidays in Europe, he would return to continue to work with me in September (before our intended departure to Europe).

July   During July, I was occupied with the transformation of the
August   Teukolsky equation in the axisymmetric case to a one-dimensional wave equation. The first steps were not difficult. But the analysis yielded one less equation than the number of functions that had been introduced. There was clearly a considerable latitude in obtaining the solution. I made the simplest assumption consistent with what I knew to be true in the Schwarzschild case and obtained a solution of the equations. But on this method, a quadrature was needed to obtain the potential. I was not too happy with this result; but was reconciled with it. My plan was to integrate this equation during September when Detweiler was to work with me. But then I had the heart attack which resulted in an immediate cancellation of our planned visit to Poland and the six months in Munich afterwards.

September   Detweiler came in early September and we went through the plan of integrating my equation to determine the "Zerilli-potential" appropriate for Teukolsky's equation in the axisymmetric case. Detweiler carried out a large number of integrations before he left for Maryland by the end of September. It was planned that he would return for a week's work in early November.

A minor piece of writing I had done earlier during the summer was an appreciation of "Marian von Smoluchowski as the founder of the physics of stochastic phenomena." This article was for a brochure which the Polish Physical Society was planning to publish in honor of its most distinguished physicist. I had written it at the request of Professor Rubinowicz.

October I was discharged from the hospital by mid-September; and during the convalescence, I decided to first write my two papers on the equations governing the Schwarzschild and the Kerr black holes; and then the joint paper on the quasi-normal modes. I wrote the first two papers during October.

Also, during October I wrote up my talk to the Innominates in April 1973, "On Some Famous Men." It was Gomer's idea that I should publish my talk in the *Bulletin of the Atomic Scientists*.

The joint paper with Detweiler on the quasi-normal modes of the Schwarzschild black hole was written during November; but it was not sent in before early December.

A surprising new turn to my ideas on the reduction of the Teukolsky equation came when Detweiler announced that by making a set of assumptions different from mine he was able to obtain an analytic solution of the basic equations that governed the trans-
November formation of the Teukolsky equation to a one-dimensional equation. The interpretation of the new solution was by no means clear since there were four possible potentials which could be complex besides. The key to the interpretation came when, during Detweiler's visit in November, the underlying relation between the Regge–Wheeler equation and the Zerilli equation became clear. We decided at this point to withdraw my earlier paper on the Kerr metric and rewrite it jointly de nova with the new point of view. Gradually during the months of November and December the whole thing fell into place; and all the requirements for the correctness and consistency of our interpretation were verified. So when Detweiler came to Chicago for
December his Christmas holidays, the problem had been fully clarified; and I could start writing the paper.

One problem to which we wanted to turn our immediate attention was to relate the different Z-functions (appropriate for the different potentials) to the perturbation of the metric coefficients. The algorithm for relating Z to the metric perturbation was clear. We foresaw some interesting possibilities and there were further problems on the horizon. With these developments in prospect, I felt moderately enthusiastic about assembling all these results for the Weyl Lectures I was to give in the fall. At long last, I seemed to have gradually emerged from my fallow period.

January 8, 1975

1971

The post-Newtonian effects of general relativity on the equilibrium of uniformly rotating bodies. V. The deformed figures of the Maclaurin spheroids (continued), *Ap. J.* **167** (1971) 447–453. March

Criterion for the instability of a uniformly rotating configuration in general relativity (S. C. and John L. Friedman), *Phys. Rev. Lett.* **26** (1971) 1047–1050. March

A limiting case of relativistic equilibrium (*General Relativity*, ed. L. O'Raifeartaigh, in honor of J. L. Synge (Clarendon Press, March 1972), pp. 185–199. The post-Newtonian effects of general relativity on the equilibrium of uniformly rotating bodies. VI. The deformed figures, of the Jacobi ellipsoids (continued), *Ap. J.* **167** (1971) 455–463. March

On the "derivation" of Einstein's field equations, *Amer. J. Phys.* **40** (1972) 224–234. August

Some elementary applications of the virial theorem to stellar dynamics (S. C. and Donna D. Elbert), *M.N.R.A.S.* **155** (1972) 435–447. September

1972

On the stability of axisymmetric systems to axisymmetric perturbations in general relativity. I. The equations governing nonstationary, stationary, and perturbed systems (S. C. and John L. Friedman), *Ap. J.* **175** (1972) 379–405. January

On the stability of axisymmetric systems to axisymmetric perturbations in general relativity. II. A criterion for the onset of instability in uniformly rotating configurations and the frequency of the fundamental mode in case of slow rotation (S. C. and John L. Friedman), *Ap. J.* **176** (1972) 745–768, revised. January

Stability of stellar configurations in general relativity, *Observatory* **92** (1972) 116–120. April

On the stability of axisymmetric systems to axisymmetric perturbations in general relativity. III. Vacuum metrics and Carter's theorem (S. C. and John L. Friedman), *Ap. J.* **177** (1972) 745–756. <u>April</u>

The increasing role of general relativity in astronomy (Halley Lecture for 1972), *Observatory* **92** (1972) 160–174. <u>May</u>

GENERAL RELATIVITY; RYERSON LECTURE; SEPARATION OF DIRAC EQUATION (JANUARY 1975–AUGUST 1977)

January 1975 The year began with writing the much postponed paper on the transformation of the Teukolsky equation for axisymmetric gravitational perturbations into the form of a one-dimensional wave equation. As I have written earlier, the key step was due to Detweiler: he first noticed how the equations governing the transformation (I had derived earlier) allowed an explicit solution; I had taken many of the key steps and carried out the required laborious reductions; but I had failed to push along the right direction. However, the analytical demonstration that all four potentials we had derived yielded the same reflection and transmission coefficients was my contribution. And Steve had verified the predictions by direct numerical integrations. So, I think it was fair that we wrote it as a joint paper. (Incidentally, the demonstration, in the joint paper on the quasi-normal modes of the Schwarzschild black hole, that the solution of Zerilli's equation (for a given $n, 1$, and σ) can be transformed to obtain a solution of the Regge–Wheeler equation was shown in the context of this paper; and it was in this manner that the equality of the reflection and transmission coefficients derived from the two equations was established for the first time.) The paper was written in January and communicated to the Royal Society. (1)

February
March
April

The months of February, March, and most of April were devoted almost entirely to preparing my Ryerson Lecture. Indeed, a larger part of the preceding months (including the weeks in the hospital in September) were spent towards the same end: I devoted, in effect, more thought, study, and effort towards preparing this lecture than to any other lecture, or course of lectures, that I have ever given. The preparation consisted in reading several biographies of Shake-speare, his sonnets (in A. L. Rowse's editions) very carefully, and listening with the text (together with Ruth and Norman Lebovitz) to all the great tragedies (in their Marlowe editions); reading several biographies of Beethoven (particularly Turner's and Sullivan's); and similarly reading several biographies of Newton; besides, the lives of Rutherford, Faraday, Michelson, Moseley, Maxwell, Einstein, Rayleigh, Abel; and books and essays by Hadamard, Poincaré and Hardy; and the works of Keats and Shelley and most particularly Shelley's *A Defense of Poetry* and King-Hele's biography of Shelley.

The Ryerson Lecture was given on April 22; and thinking back over the time and the effort I took towards its preparation, I experienced a sense of satisfaction and fulfillment which no similar effort had given me.

May
June

In early May, there was the symposium which Norman (and others) had arranged for the occasion of my 65th birthday.

It was during the symposium that I found that Steve had branched off on his own and had derived a real potential to describe electromagnetic perturbations by a method different from the one we had used. But I was anxious to go on in my direction; but Steve was clearly reluctant to go along with me: he was not committed to my point of view. And so I felt that I had to continue on my own lines; and since Steve had succeeded by his method to treat the general non-axisymmetric case, I felt the pressure on me was greater. So, after the symposium I took up the problem of electromagnetic perturbations. Restricting myself first to the axisymmetric case, I was able to obtain a very simple potential (though complex) characterizing the problem. But the question of the general non-axisymmetric case remained. Thinking of the problem one evening, the thought suddenly occurred to me that by a change of the independent variable, the Teukolsky equation for the non-axisymmetric case can be transformed into the same form as it has in the axisymmetric case. But the transformation was double-valued for $\sigma < \sigma_s$, the frequency at which super-radiance begins. In the electromagnetic case, the potential $V(r)$ in the new variable could be written down without any calculation; and the potential being real for $\sigma > \sigma_c$ $(= -m/a)$ the reason for the reflection coefficient $R \to 1$ as $\sigma \to \sigma_s$ became manifest: the potential barrier at the horizon became infinite. And the manner of onset of super-radiance for $\sigma < \sigma_s$ also became clear (though my particular manner of crossing the singularity in the potential I chose at this time had to be revised later). I completed the paper by late June and could send the paper (2) to the Royal Society before going to Varenna. On the whole, my impression is that Steve was not pleased with these alternative developments.

All during the three preceding months, there was an unhappy episode hanging over me and gnawing at me continuously. In March, I received a paper from Monique Tassoul in which she had pointed out an error in my treatment of the coupled synchronous oscillations of the Darwin ellipsoids. (The original paper was written in great rush and without the scrutiny that I should have wished: I had asked both Clement and Lebovitz to examine my analysis; but they did not have the necessary time.) Since Tassoul had made supplementary remarks wholly at variance with my own assessments of the problem, I felt that I should treat the general problem of coupled oscillations — and not only the case of synchronism — and redeem myself. I developed the necessary formulae; and left them with Norman for his checking before I left for Varenna. (And I had Donna check that my new formulae gave σ's in agreement with Tassoul's values for the synchronous case.)

Also before going to Varenna, I arranged with R. V. Jones that my article on "Of Some Famous Men," published in the *Bulletin of the Atomic Scientists*, would be reprinted in the *Notes and Records* of the Royal Society (3).

July   At Trieste and Varenna, I had long discussions with Steve with respect to our two methods. I had difficulty in convincing Steve that we had two *methods* but not two antagonistic *points of view*. I had the feeling that my approach was deeper and provided not only a unified treatment of the entire problem of Kerr perturbations but unified it also with the treatment of Schwarzschild perturbations; and Steve was unwilling to concede. I felt that the difference was one of judgment and not of substance. In any event, I was able to convince Steve sufficiently to rewrite his own paper somewhat differently; and persuade him also to generalize our earlier results on axisymmetric gravitational perturbation to cover the general case. (I think that Steve has since gradually come to my view though he has never said so.)

At Varenna, I gave the opening talk of the two week summer school that Ruffini had arranged. My talk was entitled, "Why Are the Stars as They Are?" (4) In the talk I expressed a point of view that I had long maintained but to which I had never given public expression. I was glad that at last there was a forum where I could express thoughts that had settled in my mind some forty years earlier.

August   Returning to Chicago, I had several things on my desk: the Varenna lectures, the paper on the coupled oscillations of the Darwin ellipsoids, and the preparation of the Weyl Lectures to be given in October at the Institute for Advanced Study at Princeton. And in connection with the Weyl Lectures, I felt it was important to solve the Maxwell equations in Kerr geometry completely; and I gave precedence to this over all others.

The problem of solving for the potentials describing the Maxwell field in Kerr geometry, turned out to be more subtle than I thought at first. In fact, at a certain stage I almost gave up. A new idea (Sec. 4 of the published paper) saved the situation at the "last moment."

September   I turned then to the Darwin problem. Norman had checked my analysis in the meantime, but I have the impression that he was against my publishing the paper. I felt differently: I became convinced that the anti-symmetric synchronous oscillations were more important than the symmetric oscillations that I had considered earlier and in which treatment Tassoul had found the error. Also, Tassoul's remarks at the end of her paper misrepresent, in my view, the whole situation. And I even went to the extent of saying in the introduction to my paper that what I was writing was only a "postscript" to hers. Well, I wrote the paper (5) and sent it to the *Astrophysical Journal* requesting that my paper follow Tassoul's. It did; and the unhappy episode ended. I still think that it would be useful to evaluate the σ's for the general problem solved in my paper.

I wrote my paper (6) on the "Solution of Maxwell's Equation in Kerr Geometry" soon afterwards. And I was ready for my Weyl Lectures. It was at this stage that the idea first occurred to me that I might write a book on the Schwarzschild and the Kerr Black Holes for the Clarendon Press: only the idea was forming in my mind. I made up my mind later after the Lectures.

October November

October and November were busy months. First, the Weyl Lectures and two weeks in Princeton. In November, I had trips to Cornell, Princeton, and Williams Bay (for the dedication of the new dome). Then three weeks in India.

December

Returning from India early in December, my first assignment was to write the second paper on the potential appropriate to reflection and transmission of non-axisymmetric gravitational waves by the Kerr black hole. Steve had carried out the necessary generalizations; and I had rederived them in my own way while I was at Princeton. I was able to write this paper (7) and send it to the Royal Society on January 1, 1976 ("auspiciously" as I wrote to Evans). During December, I also wrote out my second lecture at Varenna on "Linearized Perturbations of the Schwarzschild and the Kerr Metrics" (8).

January February March 1976

We spent January, February and March in Princeton; I had offices both at the Institute and at the University in Jadwin Hall. My original intention in going to Princeton was to work together with Murph Goldberger on the novel potential-barrier problems my studies on black-hole perturbations had disclosed. But Murph was too occupied with Chairman duties; and I was interested more in embarking on my long-postponed task of integrating the entire set of the Newman–Penrose equations governing the perturbations of the Kerr metric. But I had to start at the beginning.

First, I had to learn the tetrad formalism; and then "advance" to the Newman–Penrose formalism. I was not particularly enamored by the subject; but I stuck to it. Apparently none of the others I talked to (including Bill Press) had really bored into the last details: for example, I could not get anyone to tell me why there are only 18 Ricci identities.

Most of January and February were spent in working through the basics. And then I started reading the papers of Press and Teukolsky. I was, frankly, repelled by them. For example, I did not see that the authors showed any puzzlement by the separated equations not being expressed in terms of the operators $\mathcal{L}$ and $\mathcal{D}$, when the Starobinsky relations between the functions belonging to $s = +2$ and $s = -2$ were in terms of these operators. Or again, how can a relation between S_{+2} and S_{-2} involve M when the equations governing these functions do not involve it? And finally, can it really be the case that the decoupling of the equations and the separation of the variables depend on a commutation relation which stretches over several lines? I therefore started *ab initio*; and soon found that the whole analysis can be presented with simplicity and an elegance which was pleasing to me at any rate. The key relation was an elementary commutation relation involving $\mathcal{L}$ and $\mathcal{D}$.

Once I had obtained the basic equations in my forms, the Starobinsky relations for the functions belonging to $s = \pm 1$ were easily derived. But for the functions belonging to $s = \pm 2$, it was really complicated; but I did derive the radial relation between $\Delta^2 R_{+2}$ and R_{-2} I was too tired to verify the corresponding angular relation. (I verified it later after returning to Chicago. I may parenthetically add that the effort taken to derive, *ab initio*, the Starobinsky relations using my formalism were to prove immensely useful later.)

I wrote up my account of the decoupling of the equations and the separation of the variables. And since my derivations were simple, it occurred to me that I could try separating the Dirac equation which had been considered as one of the important unsolved problems of the theory. So I began with an early paper of Dirac (in the Planck Festschrift) in which he had written his equation in a tetrad frame. I took the equation as Dirac had written it and transformed it to a null-frame appropriate to the Newman–Penrose formalism. The equations I derived looked complicated. I had to leave the matter at this point since we had to return to Chicago.

But let me stop at this point and say a little about some other happenings while at Princeton. In many ways the opportunities I had to take long walks with André Weil and Freeman Dyson were among the most rewarding.

On one of my walks with André, I asked him whether his attitude to mathematics had changed over the fifty years I had known him. He replied that from his early age, he had always wished to read the classics in the originals; and that applied to Dante and Cauchy equally; and that his attitude to mathematics had always been the same as that of Hadamard who once said, "Mathematics is like ham: all of it is good — Dans le Jambon, tout est bon." But André did add that he felt that as he grew older his ability to generate fresh ideas became feebler; and on that account he himself was spending most of his time writing a history of the theory of numbers and editing papers of such great classicists as Einstein.

On another occasion Weil recalled that he was present at a reception for Einstein by Romain Roland; and that at the reception he was present during a conversation between Eli Cartan and Einstein when Cartan asked Einstein why he had not allowed for torsion in his theory; and André felt even then that Einstein did not really understand what Cartan was talking about!

I had been afraid of giving André a copy of my Ryerson Lecture. But André had heard of it and said that I was not fair in treating him that way. (But I could not tell him that it was due to my very awe of him.) So I arranged for a copy to be sent to him. After reading my Lecture, André admonished me for my reticence in not coming out explicitly with my own views. I told him that I was not sure of them anyway. And as I was leaving, Andre in his offhand manner said, "Chandra, that is a beautiful lecture you gave" — a compliment I treasure more than any "honor" I have received.

My walks with Dyson were equally rewarding. Dyson, of course, is very modest about himself. But he did say that his violin teacher was a Milton enthusiast; and that he (the teacher) persuaded him to read Milton's *Paradise Lost* when he was eight; and that he had often gone back to Milton since that time — perhaps some fifty times. (And here I am, not having read it completely from end to end even once.)

April On returning to Chicago, my first thoughts were the separation of the variables in Dirac's equations in Kerr geometry. I gave my derivation of Dirac's equations to John Friedman to check. He returned instead with a simpler derivation of Dirac's equations which made them look also simpler.

Looking at the equations, I asked incredulously, "John, are these the equations which one has failed to separate?" His answer was "yes." I assured him that I was certain they could be separated. But I could turn to them only after supper that evening (~ 8 o'clock). By ten o'clock I had separated them; and I went over the reductions carefully to check that I had not erred along the way. I had not; and I called John at about eleven o'clock (p.m.) to say that I had separated Dirac's equation. John's reply was, "I did not doubt you when you said this afternoon that you would!" And the next day I gave a seminar on my separation. I wrote the paper (9) during the following week; and sent it to the Royal Society with the request that they give the paper expeditious processing: which they did. The paper appeared in the June 15 issue.

Having solved the Dirac equation, I felt that I should complete the discussion of the two-component neutrino-equation in the manner of the other equations. The reduction to a one-dimensional wave equation was surprisingly easy; but I could not see how the absence of super-radiance was going to emerge. It was at this stage that I realized that I was not entirely correct in my formulation of the boundary conditions in my electromagnetic paper of the year before; particularly with respect to the definition of the ingoing waves at the horizon and the conditions to be satisfied across the singularity. John Friedman and (more particularly) Robert Wald were very helpful in tracing my errors. And I was able to get the cooperation of Steve in making the necessary numerical calculations; and also for providing me with an example of an integration through the singularity ... In spite of our impending departure for Cambridge, it was possible to complete the paper on the neutrino equation before leaving for England by the 27th of May.

June
July
August
On settling down in Cambridge, my first thought was to get into the problem, I had so long postponed, of the complete integration of the Newman–Penrose equations governing the gravitational perturbations of the Kerr black hole.

I first considered the linearized Bianchi identities in the gauge $\Psi_1 = \Psi_3 = 0$. These equations relate the perturbations in the tetrad with the perturbations in the spin coefficients ρ, τ, μ, π, and in Ψ_2.

These equations seemed to suggest that the perturbation $\Psi_2^{(1)}$ in Ψ_2 must vanish: otherwise, many "bizarre" identities must be satisfied; and they seemed unlikely. So I put $\Psi_2^{(1)} = 0$: in the first instance as I wrote. But very soon I decided that prudence was the better part of valour and restored $\Psi_2^{(1)}$. (Eventually, it turned out that my first suspicion about $\Psi_2^{(1)}$ was justified!) In any event, I realized that the linearized Bianchi identities related the perturbations in the spin coefficients to the perturbations in the basic tetrad. And I was at a loss to know how to proceed: the linearization of the Ricci identities (the 18 Newman–Penrose equations) seemed far too complicated to start with. But a discussion with Saul Teukolsky (who was fortunately in Cambridge at that time) directly led to the consideration of the linearized commutator relations which related the perturbations in the basic null tetrad (expressed through a matrix $\mathcal{A}$) directly to the perturbations in the spin coefficients. And it became clear that the equations to concentrate on were the 16 equations which relate the elements of $\mathcal{A}$ and $\Psi_2^{(1)}$ to the "known" perturbations in eight spin coefficients $\kappa, \sigma, \lambda, \nu$ and in ρ, τ, μ, π (the latter expressed in terms of $\mathcal{A}$ and $\Psi_2^{(1)}$ through the Bianchi identities). And since the equations relating $\mathcal{A}$ to $\kappa, \sigma, \lambda, \nu$ had the forms that the equations for the potentials of the Maxwell field that I was familiar with, I assumed that the solution must start with the solutions of these equations. And it all seemed so very easy. I was so confident of this approach at this stage (by the end of June) that I actually took "The Complete Integration of the Newman–Penrose Equations Governing the Gravitational Perturbations of the Kerr Metric" (nothing short of it!) for my talk to the Hawking Symposium in early July.

In this confident mood I went to Varenna where I gave five lectures on the perturbations of the Schwarzschild and the Kerr metrics: my talks were principally concerned only with the potential-barriers aspects of the problem though I did mention the possibility of the complete integration of the Newman–Penrose equations.

Returning to Cambridge, I began to study the various systems of equations systematically. I found that only six of the eight equations of System I (the commutator relations simplified with the aid of the linearized Bianchi identities) were linearly independent. And I thought that the remaining problem was to determine the arbitrary functions (four of them) introduced in the solution of the equations of System II (derived from the known solutions of $\kappa, \sigma, \lambda, \nu$ (At this stage, I was not very clear about the fact that the solutions for $\kappa, \sigma, \lambda, \nu$ were unspecified to the extent of the unknown relative normalization of $\Delta^2 R_{+2}$ and R_{-2}.) But the resulting equations seemed "unsolvable." I put the matter aside and tried to determine $\Psi_2^{(1)}$ — which at this stage I had left as an unknown. By a combination of various equations, I was able to show that $\Psi_2^{(1)}$ was identically zero: a surprising conclusion after some very massive reductions. (Retrospectively, this was the major result obtained in Cambridge.)

I then explored various avenues for determining the unknown functions introduced in the solutions of the equations belonging to System II. Many false trails and many misunderstandings. At one point, I thought that I had succeeded in writing Teukolsky's equations in self-adjoint form and that I had cracked the problem. With this delusion, we went on our week's holiday in the Lake district. But on the second day of our trip, I realized my mistake: which spoilt the vacation. There was nothing to do except await our return to Chicago.

eptember
October

Returning to Chicago in mid-September, I renewed my attack on the problem. I soon realized that I must abandon my method of solving the equations of System II in the manner I had solved for the vector potential for a Maxwell field: the basic problem was that there seemed to be no simple way of solving for the arbitrary functions that were introduced at that stage. So I abandoned the approach I had followed from the outset and started out afresh. (Retrospectively, I am surprised, that in spite of months of frustration that followed, the particular lines I chose at each stage among various alternatives as they appeared were exactly the right ones to have taken.)

October
November

The new method of solving the equations (giving priority to the solutions of System I) very soon led to an integrability condition. And I simply did not know how I should go about finding its implications. How indeed was one to verify immensely complicated identities involving expressions including the third derivatives of the Teukolsky functions? With the puzzle unresolved, we left for Berkeley where I was to give the Hitchcock Lectures. We visited Santa Barbara at this time and I had a chance to explain to Hartle and Chitre the nature of the problem I was knocking myself against: but they could give no assistance.

At this time Saul Teukolsky visited Chicago; and the discussions contributed to my understanding for the first time that there, underlying it all, was the question of the real and the imaginary parts of the Starobinsky constant. And since I still did not know how to verify (or deduce the consequences of) the integrability condition, I wondered, if I had perhaps, made some mistake in the massive reductions that led to integrability conditions in the forms I had derived. Fortunately, Xanthopoulos volunteered; but he had to start from the beginning and he had to be allowed at least a month.

December   During the period Xanthopoulos was checking my calculations, I decided to look into the possibility of separating the variables of the equations governing the perturbations of the Kerr–Newman black hole. At one point, I thought I had succeeded in the separation; but I realized my mistake. In any event, this interlude helped me[i] understand the meaning of the tetrad freedom; and discovering the "phantom gauge" in which Ψ_1 and Ψ_3 are reflected and transmitted by the Kerr black hole exactly as though they were electromagnetic waves. This last is a surprising fact which is not yet fully understood. Roger Penrose thinks that there is something deep here.

Xanthopoulos discovered one minor error; but it left the integrability condition in tact: it simply had to be confronted. Thus pushed against the wall, I finally realized that the only way open to me was to replace the derivatives of the Teukolsky functions (both radial and angular) in terms of the functions themselves; such replacements are possible with the aid of the Starobinsky relations. And it became clear that the end result must be the determination of the real and the imaginary parts of the Starobinsky constant as well as the relative normalization of the radial functions belonging to $s = \pm 2$.

January
1977   The New Year started with darkest gloom. The calculations that were needed to proceed with the integrability condition were long and complicated; and the hoped-for end result seemed almost hope-

---

[i]During our stay in Cambridge, we went over to London to visit Miranda Weston-Smith (E. A. Milne's grand-daughter) and her mother, Meggie Weston-Smith. Miranda was concerned that her grandfather's memory was not adequately perpetuated; and on that account she wanted to found a Milne Society. I warmly supported her initiatives and agreed to help her in any way I could.

Miranda later wrote and asked me for an "evaluation" of her grandfather which she could use for her "publicity purposes." I promised to send her such an evaluation before the "new year." So in order to keep my promise, I took the week before the new year to write an essay "Edward Arthur Milne: Recollections and Reflections" (17). I am glad I had the occasion to write this essay and pay my tribute to my first and most trusted scientific councilor: It is now deposited with the Royal Society.

less to attain. And there was the unhappy episode of the Dedekind
ellipsoids waiting in the rear all the time — more on this later. Besides, I had the weekly lectures on the "Black Hole" to give. Still,
I persisted. And to add to my nervousness, I felt my health was
deteriorating (which indeed it was).

At long last, the break occurred unexpectedly one evening when
a particular transformation involving the constants suddenly simplified the expressions and the end appeared as a distant light. Some
further work showed that the end *was* really in sight. I became so
sure that I called Xanthopoulos to say that the backbone of the
problem had been broken. But it took a week to complete all the
calculations. And after 100 pages and more of further reductions,
the real and the imaginary parts of the Starobinsky constant were
determined and so was the relative normalization constant of $\Delta^2 R_{+2}$
and R_{-2}.

February   Once the integrability condition was resolved, the completion
of the solution was accomplished with unexpected brusqueness. But
there were still a lot of ancilliary matters to wind up.

March   At this point in my lecturing, I decided to look into the matter
of the derivation of the Kerr metric. A simple and a directly verifiable
derivation was absolutely necessary if I was to proceed with the book
I had planned.

I started with the field equations as John Friedman and I had
written down in 1971. The equations took very simple forms, and
the gauge could be chosen in such a way that the occurrence of the
horizon as it does in the Kerr metric appeared most naturally. In fact,
every axisymmetric solution can be made to have a horizon exactly
as the Kerr metric. Also, a pair of real equations emerged in place of
Ernst's complex equation. It took some effort to see the various inner-
relationships in the subject. The resolution of the various problems,
as far as I was concerned, was fairly straightforward, particularly
with Xanthopoulos' help with his time and understanding.

April   With the end of the quarter, it was incumbent on me to start
May   writing the three papers: one on the derivation of the Kerr metric;

and two devoted to the integration of the Newman–Penrose equations. I decided to write two papers: one essentially presenting the derivation of Teukolsky equations and the Starobinsky relations from my point of view; and a second (and a longer paper) on the linearized Bianchi identities and the commutator relations. The paper on the Kerr metric (11) was sent to the Royal Society in mid-April; and the first (12) of the two papers on the Newman–Penrose equations two weeks later. But it took all of May to write the second paper (13).

Now I should write about the episode of the deformed figures of the Dedekind ellipsoid in the post-Newtonian approximation on which I had worked intermittently during 1972 and 1973. The principal result of the study was that the solution to the equations diverged early along the sequence; and there was no point of secular (or dynamical) instability at the point of divergence. I therefore set up the requisite fourth-order virial equations to determine the onset of the fourth harmonic instability along the sequence. I had to deal with a 16×16 matrix and all early efforts to solve it failed. But I developed a method of reducing the 16×16 matrix to a 6×6 matrix by suitable combinations of the rows and columns. Donna carried out a number of such reductions.

April
May
1976

While I was in Princeton, I gave a sample of the 6×6 matrix to Bill Press and wondered if he could find its roots. He was successful in this; and he generously solved a whole set of them. The instability did occur way down the sequence where one might have expected. I now became more uneasy about my own post-Newtonian calculations and called Tassoul in May, before we left for England, and asked him if he had ever checked my post-Newtonian calculations on the deformed figures of the Dedekind ellipsoid; and I explained my cause for uneasiness. He said that he would ask his wife to check my paper. Monique Tassoul discovered an error: the perturbed velocity field I had assumed was not general enough. But the corrections to allow for this were easily made. And Donna was to make the corrections while I was in England. On returning from Cambridge, I found that the corrections had made no difference. Then an error in applying the boundary conditions (for the boundary surface to be a streamline) was discovered. So Donna had to go through the whole calculations once more. And when in April I started writing the "Corrections and Amplifications" to the 1974 paper I found an error in one of M. Tassoul's formulae which I had tacitly assumed. And so one had to start all over again. Donna again (with unbounded patience with my errors) undertook to revise the calculations: they were completed in May.

June
July
1977
So in June after my three papers on the Kerr metric had been written, I took another week to write a "Corrections and Amplifications" to the 1974 joint paper (14). It all ended in a "comedy of errors": there was indeed no reason to have expected, in the first place, any special relationship between the Newtonian points of instability (secular or dynamical) with the point where the post-Newtonian equations governing the deformed figures of the Dedekind ellipsoid diverge — in spite of Bardeen! But there were errors of analysis in the original paper; and I am glad that they have been corrected at long last.

Thus by the end of June, all of the four papers which were a terrible burden and a continuous nervous drain on me all during the fall and winter were at last written and sent for publication.

In July, I wrote out my conversations with Hardy and Little-wood relating to Ramanujan's attempt to commit suicide. So much perfumed nonsense has been written about Ramanujan that I felt some record of this tragic aspect in the life of Ramanujan should be deposited with the Archives of the Royal Society. I also included an account of my discovery of the one and only authentic photograph of Ramanujan. Both these accounts are now with the Royal Society (15, 16).

I also wrote my long postponed introductory part of the joint review with N. Swerdlow of the three Neugebauer volumes (18).

In August we went to GR8 in Waterloo, Canada, where I gave a talk on the Kerr metric. On our return I had my long post-poned aorto-catherization; it indicated heart surgery (scheduled for tomorrow). So this period of interest in the Schwarzschild and Kerr black holes ended as it began — in a hospital.

Billings Hospital, August 24, 1977

References

(1) On the equations governing the axisymmetric perturbations of the Kerr black hole (S. C. and S. Detweiler), *Proc. Roy. Soc.* **345** (1975) 145–167; (February 1975) RYERSON LECTURE, Shakespeare, Newton, and Beethoven or patterns of creativity, delivered April 22, 1975.

(2) On a transformation of Teukolsky's equation and the electromagnetic perturbations of the Kerr black hole, *Proc. Roy. Soc.* **348** (1976) 39–55. (July 11, 1975)

(3) Verifying the theory of relativity, Notes and Records, *Roy. Soc.* **30** (1976) 249–260. (June 3, 1975)

(4) Why are the stars as they are? Varenna Lectures. (September 23, 1975)

(5) On coupled second-harmonic oscillations of the congruent Darwin ellipsoids, *Ap. J.* **202** (1975) 809–814. (September 11, 1975)

(6) The solution of Maxwell's equations in Kerr geometry, *Proc. Roy. Soc.* **349** (1976) 1–8. (September 15, 1975)

(7) On the equations governing the gravitational perturbations of the Kerr black hole (S. C. and S. Detweiler), *Proc. Roy. Soc.* **350** (1976) 165–174. (January 1, 1976)

(8) On the linear perturbations of the Schwarzschild and the Kerr black holes, Varenna Lectures. (December 19, 1975)

(9) The solution of Dirac's equation in Kerr geometry, *Proc. Roy. Soc.* **349** (1976) 571–575. (April 21, 1976)

(10) On the reflexion and transmission of neutrino waves by a Kerr black hole (S. C. and S. Detweiler), *Proc. Roy. Soc.* **352** (1977) 325–338. (June 2, 1976)

(11) The Kerr metric and stationary axisymmetric gravitational fields, *Proc. Roy. Soc.* (April 15, 1977)

(12) The gravitational perturbations of a Kerr black hole. I. The perturbations in the quantities which vanish in the stationary state, *Proc. Roy. Soc.* (April 27, 1977)

(13) The gravitational perturbations of the Kerr black hole. II. The perturbations in the quantities which are finite in the stationary state, *Proc. Roy. Soc.* (June 17, 1977)

(14) The deformed figures of the Dedekind ellipsoids in the post-Newtonian approximation to general relativity; corrections and amplifications (S. C. and Donna Elbert), *Ap. J.* (July 6, 1977)

(15) An incident in the life of S. Ramanujan, F.R.S.: Conversations with G. H.

Hardy, F.R.S. and J. E. Littlewood, F.R.S.; and their sequel (Archives of the Royal Society).

(16) On the discovery of the enclosed photograph of S. Ramanujan, F.R.S. (Archives of the Royal Society).

(17) Edward Arthur Milne: Recollections and reflections (Archives of the Royal Society).

(18) Book review: *A History of Ancient Mathematical Astronomy* (3 volumes) by O. Neugebauer, *Bulletin of the American Mathematical Society.*

(The dates in parentheses are the dates when the respective papers were submitted for publication.)

General Relativity; Kerr–Newman Perturbations (August 1977–December 1978)

During the two months, following my surgery, I was, physically, in a state too painful to do any serious scientific work. By November, I had sufficiently recovered to plan a little for the book on the Schwarzschild and the Kerr metrics that was constantly at the back of my mind. And with the book in view, I began studying Lovelock and Rund on Differential Geometry as a preparation towards the first chapters that were to be devoted to the Cartan calculus. But my feeling of assurance at this time, that my major investigation related to the solution of the Newman–Penrose equations had been completed, was soon to receive a serious setback.

Sometime in November, John Friedman visited me during my convalescence; and among other things, he expressed the view that my solution for Ψ, in terms of indefinite integrals over the Teukolsky functions, left the subject in a state of incompleteness: he felt that it should be possible to express Ψ explicitly in terms of the Teukolsky functions. This remark of John's touched a sensitive spot: I had in fact been dissatisfied with the appearance of indefinite integrals in the solution for Ψ: it had left in total obscurity the meaning of the two differential equations for Ψ (II, Eqs. (103) and (104)).[j] I was very piqued with myself when John left.

Later that evening, I sat at my desk and turned over the sheets of my detailed calculations pertaining to Ψ. And looking at the solution for Ψ as I had left it (II, Eq. (96)), I realized that the finiteness of Ψ for $a \to 0$ — a requirement I had not considered but should have chosen the positive sign for C_1 so that $(C_1 - \Gamma_1)/a$ was finite in the limit $a = 0$. Writing the coefficient $(C_1 - \Gamma_1)/a$ as $(C_1^2 - \Gamma_1^2)/a(C_1 + \Gamma_1)$, the separability of Ψ as a product of a function of r and a function θ became immediately manifest. This was an unexpected and a total surprise.

With Ψ expressed as a constant $\times \mathcal{R}(r)\zeta(\theta)$ the fact that there are two equivalent forms for $\mathcal{R}$ and ζ soon became evident. The alternative expressions implied the existence of identities among the Teukolsky functions, identities I could not imagine how one could verify *ab initio*. The meaning of the differential equations for Ψ (II, Eqs. (103) and (104)) as differential equations for $\mathcal{R}$ and ζ also became clear.

All of the foregoing reductions were carried out during the last weeks of November and the early part of December. But I felt uneasy about the identities: they were the results of massive reductions. I felt that a direct numerical verification of the identities will provide assurance that no errors had crept into the reductions. With this feeling, I called Detweiler and asked him if he could provide tables of the Teukolsky functions in my normalizations: the radial functions X and Y as I had defined them and the angular functions S_{+2} and S_{-2} normalized in the standard way.

I also expressed the need to verify my formulae relating the derivatives of the Teukolsky functions to the functions themselves. Detweiler agreed to provide me with the requisite integrations; but he thought that he may not be able to get round to them before another month or six weeks.

---

[j]The reference here (and in the sequel) is to my paper, *Proc. Roy. Soc.* **358** (1978) 421.

In some ways, I welcomed the period of waiting: I had promised Israel that I would send him my chapter on the Kerr metric for the Einstein Centennial Volume by the end of November. I had requested and obtained the postponement of the deadline to the end of January on account of my illness. And the writing of the article could not be delayed any longer. By this time, I was getting back to my normal schedule; and the whole of January was spent in writing the article. It was completed and sent by the end of the first week of February.

In writing the article, I had to collate much new material: an account of the tetrad formalism, geometrical considerations pertaining to the Newman–Penrose formalism, and a unified treatment of the potential-barrier problem.

In February, while still waiting for Detweiler's integrations, I began to think once again about the problem of the Kerr–Newman perturbations: a problem I had attempted and left without success a year earlier. I tried various approaches; but in vain. It then occurred to me to look into what had been done with respect to the Reissner–Nordström black hole. I was surprised to find that the Newman–Penrose equations appropriate for the Reissner–Nordström perturbations had not even been decoupled; and further that Moncrief (who had obtained decoupled one-dimensional wave equations for both parities) did not seem to know that the solutions for the two parities should be related and that a separate discussion of the stability for the two parities was unnecessary. I gave a Friday seminar on these matters; and when I expressed these views, Ashtekar and Bob Wald (?) questioned me as to how I could be certain that what applied to Schwarzschild applied also to Reissner–Nordström. My reply was, that if my demonstration for the simple relation between the solutions for $z^{(+)}$ and $z^{(-)}$ in the case of the Schwarzschild perturbations was "obvious" — as several, including Hawking had claimed — then why should it not be equally obvious for the Reissner–Nordström perturbations. But at this stage, I had not decoupled the appropriate Newman–Penrose equations.

I cannot quite recall how it happened. Trying idly a certain combination of the Weyl scalars and the spin coefficients, I was taken aback when I found that the combinations I had chosen did in fact decouple the Newman–Penrose equations. The important point here is that the gauge I had considered as "god-given" (namely, one in which the Maxwell scalars ϕ_0 and ϕ_2 are identically zero) is the one which decouples the equations. The "phantom gauge" (as Roger Penrose had described it) had proved its usefulness.

Once the Newman–Penrose equations had been decoupled, the reduction of the decoupled equation to the form of a one-dimensional wave equation could be effected by the transformations used in other contexts. I soon verified that the transformation — to Moncrief's odd-parity equations could be effected with the assumptions β = constant and $f = 1$, assumptions which were valid in the Schwarzschild case. But I was blocked for some time as to how to deduce the even parity equations: Moncrief's solutions were far too complicated. It was only a few days later that I realized that the even parity equations will automatically follow from the dual transformation with β having the negative of the value appropriate for the odd-parity transformation. Once this fact became clear the relation between $z_i^{(+)}$ and $z_i^{(-)}$ could be deduced — and Ashtekar and Wald had to eat "crow."

All of the foregoing was done during the early weeks of March. One question remained unclarified: how is one to relate the solutions $z_i^{(+)}$ and $z_i^{(-)}$ to the electromagnetic and the gravitational fluxes of the incident and reflected radiations. (The question was clarified only in June when I had the occasion to discuss the problem with Matzner in Austin, Texas.)

Meantime, Detweiler had sent his integrations for X, Y, S_{+2} and S_{-2} having verified my formulae for the derivatives. Donna verified my identities (but discovered that there was an error of a sign in one of the equations for $\mathcal{R}$ and ζ in terms of X, Y, S_{+2} and S_{-2}).

A further fact of considerable importance which emerged during this period was the discovery of an error of a factor 2 in a key equation of mine which had led me to the result that the perturbation, $\Psi_2^{(1)}$, in the Weyl scalar, Ψ_2, is zero. John Friedman was considerably disturbed that one could deduce this result. He argued, correctly, that while $\Psi_2^{(1)}$ can be set equal to zero (using two of the four co-ordinate degrees of freedom), one should not be able to deduce that it is zero in a gauge invariant theory such as the Newman–Penrose formalism. He had accordingly asked his student, Eliane Lessner to check through my calculations. Mrs. Lessner found the error; but its disclosure was due to Friedman's insistence. At this stage I did not realize the full implications of this result: while I was naturally disappointed that I had made the error, the fact that it did not really matter ($\Psi_2^{(1)}$ can after all be set equal to zero) left me, up to a point, undisturbed.

It was now getting towards the end of March and it was high time that I turned my thoughts towards the course on Cosmology that I had originally scheduled for the winter quarter and which I had postponed to the spring quarter. (My intention in wanting to give a course on Cosmology was largely to get acquainted with the subject with a remote idea that I might turn to it once my interest in the black hole solutions had tapered off and also because cosmology is the one area of astronomy in which I had taken no serious interest.)

I had decided that my course would be devoted to the Russian work on the Bianchi models and the perturbations of the Friedman models. The study of the Russian work turned out to be far more time consuming than I had anticipated. During all of April and May I could not find time for anything else.

By early June, my thoughts turned to writing up my two papers: on the separability of Ψ and the resulting identities and my decoupling of the Newman–Penrose equations describing the Reissner–Nordström perturbations. But first, I had to go to Austin, Texas to give a Schild Memorial Lecture. In many ways this visit to Texas was a fortunate one. I had the occasion to meet and discuss with Richard Matzner my results on the Reissner-Nordström perturbations. And I found that in one of his papers, Matzner had obtained the relation that I was looking for: the relation of my functions $z_i^{(\pm)}$ to the fluxes in the incident electromagnetic and gravitational waves. It was clear to me at once that with Matzner's relation I could readily ascertain how an arbitrary superposition of incident electromagnetic and gravitational waves will be reflected and transmitted by the Reissner–Nordström black hole. I completed the solution on my return to Chicago.

The completion of the theory of the perturbations of the Reissner–Nordström black hole via the Newman–Penrose formalism suggested to me that I should work out the metric perturbations along the lines of my earlier treatment of the Schwarzschild perturbations. Since I had been discussing my work with Xanthopoulos all along, I asked him whether he would consider staying in Chicago for some $2\frac{1}{2}$ months after taking his degree and collaborate with me on this aspect of the problem. He agreed; but he asked me what the point was in doing the metric perturbations in my alternative way when the problem had been "solved" by Moncrief by a different method. I said that I wanted the subject to have an architectural unity; and Moncrief's method was simply out of place in my structure. Xanthopoulos' response was that I could afford to take such a point of view, meaning, apparently, that young men in the beginning of their careers could not afford to take a similar outlook.

With the belief that the investigations appropriate to my two papers were completed, I set about writing my third paper on the Kerr perturbations. But doubts, vaguely entertained, emerged with insistent force, I had been aware all along that I had ten degrees of gauge freedom: six from the choice of the tetrad frame and four from the general covariance of the theory. I had used up four of the six tetrad freedoms to set Ψ_1 and Ψ_3 equal to zero. And setting $\Psi_2^{(1)}$ equal to zero exhausted two of the four coordinate degree freedoms. And the fact that the diagonal elements A_1^1, A_2^2, A_3^3 and A_4^4 of the matrix A were left unspecified, meant that there was no freedom left to let one of $F + G$ and $J - H$ unspecified. I discussed this matter, in these terms, with John Friedman one morning (he was spending the summer months at Chicago); and it became abundantly clear that a further relation between $F + G$ and $J - H$ had to be found supplementing the information already obtained via the solution for Ψ. And the question was how?

Since I had satisfied all of the Bianchi identities and the commutation relations, I had to go to the Ricci identities, several of which had already been verified. A careful re-examination of the Ricci identities convinced me that I should consider the Newman–Penrose equations (4.2a), n, g, and p which involve the derivatives $(\partial^* k, \partial \nu, \mathcal{D}\lambda, \Delta\sigma)$ combined, respectively, with the derivatives $(\mathcal{D}\rho, \Delta\mu, \partial^*\pi, \partial\tau)$. And I recalled that I had in fact linearized these equations — and indeed, for the same purpose! — two years earlier at Cambridge. This was the first of the "lucky breaks." But there is no gainsaying that I was extremely discouraged at this stage and during the subsequent weeks.

The systematic reduction of the chosen Ricci identities was a disheartening matter most of the way; but it was punctuated by further "lucky breaks" which made the work possible.

In the reductions, it was necessary to keep the prime objective always in focus. The objective was to obtain an equation for $z_1 - z_2$ where $z_1 = K(J - H)\cos\theta$ and $z_2 = -irQ(F + G)\sin\theta$, since $\mathbf{Z} = z_1 + z_2$ is already known to be a constant $\times \mathcal{R}\zeta$. It was equally important to have devised a "bracket" notation which enabled me to write the various complicated relations and identities in manageable forms. In reducing the equations, several identities appeared along the way; and their verification gave insight into the various quantities which emerged from the analysis. In some ways the elementary identities among the coefficients A_1, A_2, B_1, B_2, and E which appeared in the four equations for Z_1 and Z_2 were unexpected. They emerged only slowly; but retrospectively they seem to be contrived exactly for the equations to be solvable. When finally an integrability condition for the existence of a solution for $Z_1 - Z_2$ emerged, I had the vision of some 200 pages of calculations similar to those required to consider the corresponding integrability condition for Ψ. Some two months of work appeared in store and I was almost inclined to leave the matter at this stage since the problem had after all been solved in principle. But my unwillingness to accept defeat turned out to be fortunate. By an extraordinary piece of luck, the calculations turned out to be relatively easy: my familiarity with the identities involving $\mathcal{R}$ and ζ and the adaptability of my bracket notation were happily what were needed. And so the problem got solved: but one lacuna still remains which I must eventually fill.

The emergence of so many identities among the Teukolsky functions — both radial and angular — was a most unexpected outcome. As Friedman and Teukolsky later described (independently), the identities are "astonishing."

By the third week of July, I was at last ready to start writing my two papers ((1) and (2)). With great effort, it was possible to write them up and mail them to the Royal Society early during the second week of August (14th). Only a few days remained to prepare the "Oppenheimer Lecture" (3) that was to be given in Los Alamos on 17th August.

Returning from Los Alamos, I had to prepare the address that I was to give in Rome at a symposium in honor of Amaldi. I wrote out my lecture; but my trip was aborted: I had not noticed that my passport had expired. But then I had a few more days to write the first part of my promised article on "General Relativity and Cosmology" for *The Great Ideas Today*. I had promised the article a year earlier: but my illness and the unexpected stumbling blocks in completing my third paper on the Kerr perturbations prevented my writing it in time. In fact, September 20 was a postponed deadline for the delivery of the first part of the manuscript on "Relativity."

The writing of the article required considerable concentration: I wanted to write one of which I would not be ashamed. With considerable effort, I did manage to send the article to *The Great Ideas Today* (4) just a day or two before our departure to Santa Barbara on September 22.

By this time Xanthopoulos had completed his reduction of the metric perturbations of the Reissner–Nordström black hole, and had derived *ab initio* both the odd and the even-parity one-dimensional wave-equations following the methods of my treatment of Schwarzschild perturbations.

The first week in Santa Barbara was a period of adjustment. Among other things, I had to plan a course of lectures on the "Schwarzschild and the Kerr Metrics" that I was scheduled to give. (Altogether I gave 20 lectures devoted, almost entirely, to my own way of looking at the matter.)

By the end of the first week, I decided that my first task must be to write my joint paper with Xanthopoulos on the metric pertur-bations of the Reissner–Nordström black hole. I found that I had, effectively, to work through the entire problem *ab initio* though the fact that Xanthopoulos had identified the functions $z_i^{(\pm)}$ with certain combinations of the metric perturbations made the work essentially "routine." But I did obtain the general form of Maxwell's equations in a non-stationary axisymmetric space-time and found that their linearization about the Reissner–Nordström solution gave the requi-site equations more directly. Also, since Xanthopoulos had given so few details in his notes, my reductions were independent and differ-ent from his in essential details. I am glad I did the entire reductions myself since I was enabled to write the paper in my own way. The reductions took me about two weeks; and it took another week to write the paper. I sent the $(n-1)$ draft to Xanthopoulos for ap-proval; but I wrote the nth copy without waiting for his comments. He did have some suggestions which I incorporated in the final copy. The paper was sent in to the Royal Society during the third week of October (5).

Already during the week when the paper for the Royal Society was being typed, I had started working on the second part, on Cosmology, of my article for *The Great Ideas Today*. This part was more difficult to write than the first part on Relativity. I had to read and digest a lot of new information and develop my own point of view. Discussions with J. Hartle and B. Hu were very helpful.

I was able to give the completed manuscript for typing before I left for Charlottesville on November 7 where I was scheduled to give the Karl Jansky Lecture on November 8. (The Lecture was a disappointment for me, though the audience, Roberts, and Hogg seemed satisfied.) On returning to Santa Barbara, I was able to read and correct the typed manuscript on Cosmology. I sent it to Van Doren on November 13 as I had promised (4).

At long last, I was relieved of the constant pressure under which I had been working since December 1977.

After a week of relaxation, I began to think about the future. And immediately the problem of the separation of the variables and the decoupling of the Newman–Penrose equations governing the perturbations of the Kerr–Newman black hole emerged once again from the shadows to which I had consigned it in March. It was frustrating to realize that all my experience with the Kerr, the Schwarzschild, and the Reissner–Nordström black holes were of no avail in the Kerr–Newman context. Essentially some new ideas were needed; and what I needed before all else was the freedom to relax and contemplate with no pressure. I had hoped for these at Santa Barbara; but the pressure of unfulfilled tasks did not permit such pleasures.

Looking back over the past three years (after our return from India in December 1975), I find that I have been taxed and continuously burdened. Indeed, I cannot recall that I had been so subjected at any earlier period in all my 50 years of scientific life . . .

Tomorrow we will return to Chicago; and my book loomed ahead.

December 1, 1978
Santa Barbara

References

(1) The gravitational perturbations of the Kerr black hole. III. Further amplifications.
(2) On the equations governing the perturbations of the Reissner–Nordström black hole.
(3) Einstein and general relativity — historical perspectives.
(4) Einstein's general theory of relativity and cosmology.
(5) On the metric perturbations of the Reissner–Nordström black hole.

1979 — A Year of Failures and of Obligations

It is now almost exactly a year since we returned from Santa Barbara. And I had high hopes for 1979: hopes for separating and decoupling the Newman–Penrose equations governing the perturbations of the Kerr–Newman black hole and for making some progress with my book. But not only were those hopes shattered, everything else I attempted also turned to failures. The past year has, in fact, been the most unsuccessful of all the fifty years of my life as a scientist. How did this come to pass?

On returning from Santa Barbara, my first thoughts were naturally directed to the problem of the perturbations of the Kerr–Newman metric. As I said I had high hopes. I reasoned as follows: If the Newman–Penrose equations governing the perturbations of the Kerr–Newman metric, in the forms and in the gauge I had written them down in my first paper on the Kerr perturbations, should separate and decouple, then, if in the analysis leading to this separation and decoupling, we set $Q_* = 0$, it must reduce to my analysis for the Kerr metric, while if we set $a = 0$, it must reduce to my analysis for the Reissner–Nordström metric. I, therefore, tried to amalgamate the ideas and the procedures I had learnt in the two cases. I tried all possible combinations; but they all led to blind alleys.

I abandoned my efforts on this problem by mid-January since I could no longer postpone writing my promised contribution to the Schild–Memorial volume. However, I decided not to give a general account of *The Potential Barriers Around Black Holes* (the title of the lecture I had given, in 1978) but rather give an *ab initio* account of the Schwarzschild perturbations from the vantage point I had gained over the years, particularly, after my treatment of the Reissner–Nordström perturbations. Small as this effort was, it took up most of February. But it was a useful exercise to have undertaken: it rekindled my interest in the many unresolved questions of the theory still waiting for clarification.

During the month of March, I returned once more to the Kerr–Newman perturbations with some further combinations of the old ideas; but to no avail. And going to Princeton for the Einstein celebrations provided the occasion for a final break with this problem.

Returning from Princeton, I had to think in earnest about the lecture on 'Beauty and the Quest for Beauty in Science' that I had agreed to give at the one-day symposium in honor of Robert Wilson that Jim Cronin was organizing. The preparation for this lecture required much concentrated thinking; and several aspects of 'beauty' that I had barely considered in my Ryerson Lecture, had to be thought afresh. The lecture was eventually published in *Physics Today*; but I had to insist that it was published exactly as I had written and without any changes.

In May, I returned to the unresolved questions on the theory of the Schwarzschild perturbations: questions that had long puzzled me and more insistently since writing my review for the Schild–Memorial volume. The principal question concerned the symmetry of the equations relating the solutions for $z^{(+)}$ and $z^{(-)}$, appropriate for the perturbations belonging to opposite parities, and the lack of any corresponding symmetry in the equations governing $z^{(+)}$ and $z^{(-)}$. A related question was: could one have foretold that an explicit relation between $z^{(+)}$ and $z^{(-)}$ exists (apart from its symmetry) from an examination of the governing equations? A third question was the reason for the reduction of the three first-order equations, governing the three scalars determining the metric perturbations, to a single second-order equation (see Chandrasekhar, *Proc. Roy. Soc.* **343** (1975) 289, Eqs. (27)–(29)). I began serious consultations with R. Narasimhan on these questions at this time. I had in fact, briefly talked to Narasimhan about them a year earlier. After our return from Santa Barbara, Norman Lebovitz told me that Narasimhan had talked to him about my questions and had indicated that they could all be resolved very simply by considering the equations in the complex plane and by investigating the behavior of the solutions at their singular points. With the disappointment over my lack of success with the Kerr–Newman perturbations, the possibility of a 'breakthrough' in these old unresolved questions was exciting and I turned to them with enthusiasm.

Narasimhan's view was the following: The equations for $z^{(+)}$, $z^{(-)}$ and Y must have the same monodromic group; and from the fact that they do, one can infer that each of them can be expressed in terms of one or the other of the remaining two. He further stated that one should be able to relate any one of them with the solution of an equation which had the same monodromic group but with all its singular points regular. The last of these statements implied, in fact, that $z^{(+)}$ and $z^{(-)}$ can be expressed in terms of some hypergeometric function. If this last could be accomplished, it would indeed be a most exciting outcome. And during the next few weeks, I learnt about the monodromic group from the books of Ince and Poole.

There was no difficulty in showing that the equations for $z^{(+)}$ and $z^{(-)}$ do indeed have the same monodromic group. The equation with the same monodromic group but with regular singular points could quite easily be constructed. The solution of this last equation was readily expressible in terms of the incomplete Beta function. But I could see no way in which the solution of this last equation could be related to $z^{(+)}$ and $z^{(-)}$. In fact, together with R. Sorkin, I concluded that Narasimhan must be under some total misapprehension of what it is I was seeking. My efforts via the monodromic group thus collapsed.

At about the time I was following the monodromy trail, Xanthopoulos, during a visit to Chicago, told me about his efforts to complete the solution for the metric perturbations of the Reissner–Nordström black hole. It became clear to me that in the course of his work, Xanthopoulos had, in fact, discovered a special integral of the basic radial equations. In particular, the three equations for the radial functions, I had derived in the context of the Schwarzschild black hole, allow a special integral. This fact made it immediately clear why the three equations are reducible to a single second-order equation. The last of the three questions I had puzzled over had thus found its answer. But the principal questions remained unanswered. The answers to them came very unexpectedly.

I cannot quite recall how I came to trying the particular sequence of transformations which resolved the basic questions. I seemed to have hit upon them by accident while turning over in my mind the equations of the transformation theory as I had set it out in the appendix of my paper on the Reissner–Nordström perturbations. In any event, by explicitly evaluating the expressions for $V^{(\pm)}$ given by the theory and making use of the nonlinear differential equation for "F", I was able to show that the potentials, $V^{(\pm)}$ for both the Schwarzschild and the Reissner–Nordström black holes, are included in the general forms,

$$V^{(\pm)} = \pm\beta\frac{df}{dx} + \beta^2 f + kf \,,$$

where β and k are constants and f is an arbitrary continuous function which together with all its derivatives have bounded integrals over the range $(+\infty, -\infty)$. By showing that $V^{(\pm)}$ are of the foregoing form, I had quite inadvertently resolved the two questions which had puzzled me since 1974. And further, by making use of the general forms for $V^{(+)}$ and $V^{(-)}$, I was able to establish the infinite hierarchy of the integral equalities between them. I had suspected the existence of this hierarchy of integrals after my conversations with Ken Case following one of my Weyl Lectures in Princeton in 1975. The final resolution of these long standing questions is the one bright spot in an otherwise bleak canvas.

By the third week of July, I was able to write my paper 'On One-Dimensional Potential Barriers having Equal Reflexion and Transmission Coefficients' and send it to the Royal Society. During August, Donna evaluated explicitly the first five integrals of the hierarchy for the Schwarzschild potentials. I was glad that the results of her evaluation could be included in the paper: it provided me the opportunity to make one last acknowledgment to her assistance over the years.

With the end of July approaching, I had to turn my thoughts, earnestly and seriously, to the invited discourse on 'The Role of General Relativity in Astronomy — Retrospect and Prospect' I was to give at the I.A.U. meeting in Montreal in August. Since this was my first attendance at a meeting of the I.A.U. after the one in Paris in 1934, I wanted the discourse to reflect my views and my attitudes to research in theoretical astronomy. I wanted it, in fact, to be in some ways my astronomical testament.

As it turned out, my lecture had a very large audience with a large overflow and with no room even for standing. For my part, I do not have any feelings as to how effective my lecture was. But Martin thought it was 'most courageous'.

Returning from Montreal, there were a number of things to attend preparatory to Donna's departure to 'greener pastures'. And then we went on a short holiday to the Lake Superior region during the second week of September.

After the holiday, I was not free to return either to research or to my book. Instead, I had to write my contribution to a *General History of Astronomy* which Owen Gingerich was editing. My contribution was to be on 'General Relativity: The First Thirty Years'. It was a weak moment in 1977 that I had given in to Gingerich's repeated entreaties. I now regretted my weakness even more. For, after spending all of October in writing the essay, Gingerich wanted me to condense the article and rewrite parts of it at a less technical level. I simply refused: I had already spent more time on it than I could afford. And the matter stands there now.[k]

---

[k]The article will now be published in *Contemporary Physics*.

And still, I was not free. I had to prepare the text of the Milne Lecture I am to give in Oxford on December 6. The Lecture was to be on 'Edward Arthur Milne: His Part in the Development of Modern Astrophysics'. I was not at all sure how my assessment of Milne's work would be taken by others, especially his family. I was convinced that the work of a scientist must stand on its own; and no assessment is of value if it is not totally honest. And so the year ended with freedom at last to start on my book. I hope that a year from now I shall have a gladder tale to tell.

References

(1) On the potential barriers surrounding the Schwarzschild black hole (*The Alfred Schild Memorial Lecture Series*).
(2) Beauty and the quest for beauty in science (*Physics Today*).
(3) On one dimensional potential barriers having equal reflexion and transmission coefficients (*Proc. Roy. Soc.*).
(4) The role of general relativity in astronomy — retrospect and prospect (IAU – Discourse).
(5) The general theory of relativity — the first thirty years (*Contemporary Physics*).
(6) Edward Arthur Milne: His part in the development of modern astrophysics (*Quarterly Journal of Roy. Astron. Soc.*).

November 26, 1979

Postscript

The week following November 26 was the week of the visitors — Bondi and Penrose — and the week of the Christmas Lectures. I was glad that I had introduced Bondi both at the Physics Colloquium on Thursday (November 29) and at the Christmas Lecture on Friday. I did not get to see Penrose very much during the week even though we did have him at home for both a dinner and a breakfast. The week was a hectic one; and I was glad that I had already written the complete text of my Milne Lecture. Penrose left on Sunday; and we left on Monday (December 3).

We arrived in Oxford late in the afternoon of Tuesday. We rested and then went to visit Penrose at the Mathematical Institute. On Wednesday, I gave a talk at Penrose's Seminar about some aspects of my work. It was a pleasure to talk to an audience familiar with my work and appreciative of it. Roger was exceptionally delightful in what he said. It is only rarely that I have had such pleasurable occasions during these latter years.

On Thursday, I gave my Milne Lecture. I did not get any feeling at the time as to how effective my Lecture was. But Roger wrote handsomely afterwards.

In London, we stayed in the rooms of the Royal Society; and we had a very pleasant evening at the Seatons. Our stay in Cambridge was memorable only in our visit to Kate and David Shoenberg. We were also entertained by Martin Rees, Lynden Bell, Nigel Weiss and Allan Hodgkin (at the Lodge).

Returning from England, I turned my thoughts to the book. Instantly, I felt uneasy about the single lacuna in the information still left at the end of Paper III; and I was not happy with what appeared to me as a clumsy finale. The idea of reproducing such an inelegant solution in my book was distasteful. I began wondering whether I should not after all reduce the six Ricci identities involving the derivatives of α, β, γ and ϵ. And, I began to be concerned equally with the form of the solutions, particularly of $\mathcal{R}$ and ζ, in the Schwarzschild limit, $a \to 0$. While casually turning over the accumulated sheets of my notes, it suddenly occurred to me that I should perhaps first consider the pair of equations, derived from the same Ricci identities, but complementary to the one that I had considered in Paper III. Once the thought had occurred, I could not doubt the outcome. Indeed, the steps that I had to follow to obtain the equations that will determine z_1 and z_2 explicitly, were very clear.

I made the principal reductions during the Christmas recess (December 26). A few days later, I had completed the necessary reductions and I had in front of me explicit formulae for z_1 and z_2. And the fact that $z_1 + z_2$ must equal Ψ required that an identity be satisfied. I naturally wanted to verify the identity numerically, but the numerical test failed: there was clearly some error that had crept into the analysis. As repeated checkings did not disclose any error, I began even to doubt the entire procedure. But a close discussion with John Friedman dispelled any such doubts. It was, nevertheless, very unsatisfactory to have to leave for Paris on January 20 with the discrepancy still unresolved. And there was also the manuscript of my Unesco Lecture to get written up and typed before leaving.

The week in Paris was uneventful except for a day I had with Carter at the Meudon Observatory and the chance encounter with the Spitzers, Doreen materializing from nowhere a few minutes before my Lecture.

Returning from Paris, I continued to be concerned with my failure with the identity. And again, by great good fortune, while turning over once more the pages of my notes in an attempt to check my formulae for z_1 and z_2 in the limit $a \to 0$, I was startled to find that one term was dimensionally wrong: by an error in transcription, I had omitted a factor $1/r$ in one of the terms. This was on the afternoon of Sunday, February 3. I at once went to the Institute; and within an hour the discrepancy that had worried me for a month was gone! By February 6, I had tied the various loose ends including the verification of the identity involving z_1 and z_2 for the Schwarzschild limit.

I was able to write the last (and the fourth) paper of my series on the solution of the gravitational perturbations of the Kerr black hole during the following two weeks; and on February 19, the paper was mailed to the Royal Society. And as I wrote on the fly leaf of the preprint of the paper to John Friedman and Saul Teukolsky,

At long last! But meantime, I seem to have let life pass me by!

In any event, so far as I could see, there was no further obstacle that would prevent me from starting on my book. And I shall return to these pages when and if I had written the book.

References

(1) Black holes: the why and the wherefore (Unesco Lecture).
(2) The gravitational perturbations of the Kerr black hole: IV. The completion of the solution (*Proc. Roy. Soc.*).

February 21, 1980

1980, 1981: THE MATHEMATICAL THEORY OF BLACK HOLES

One of the requirements, associated with the Hermann Weyl Lectures I was to give at the Institute for Advanced Study in Princeton in October 1975, was that the lectures should be published. It occurred to me then that I might expand the lectures[1] into a full-scale book. And when writing to Denys Wilkinson on October 14, recommending Parker's Cosmical Magnetic Fields for the International Monographs in Physics, I inquired how he might respond to a book of my own on the Schwarzschild and the Kerr metrics. Wilkinson responded enthusiastically; and I wrote a more detailed letter with an outline of the proposed book (see p. 157) on October 21; and a formal contract with the Clarendon Press was signed on December 1, 1975.

At the time the contract was signed, only six of the sixteen papers (that were eventually to appear in the *Proceedings of the Royal Society*) had been published. The entire work on the gravitational perturbations of the Kerr black hole, the separation of Dirac's equation in Kerr geometry, as well as the theory of perturbations of the Reissner–Nordström space-time, were all in the future. A comparison

---

[1]The expansion of my Weyl lectures was actually published in Hawking and Israel's volume. *General Relativity — An Einstein Centenary Survey* (Cambridge, England, 1979), Chapter 7, pp. 371–91.

of the 'outline' given by Wilkinson in October 1975 with the final contents as it emerged in January 1982 will show that almost 75% of the completed book was yet to be investigated when the idea of the book occurred.

As it turned out, the last of my sixteen papers in the *Proceedings of the Royal Society* was communicated only on February 21, 1980. I started on the book on March 1, 1980; and it was to fully occupy me for almost two years.

Let it suffice to say here that the manuscript for the first nine chapters (together with accompanying illustrations) were handed over to Mr. Manger at the Kennedy Airport (New York) on September 19, 1981 prior to our departure to Poland. Chapter X was completed and sent on November 16. And all of Chapter XI (exclusive of the last section 114), the Appendix, and the Epilogue, were sent on January 7, 1982. Finally, on January 26, the last section 114 was sent. And on Friday, January 29, a call from Mr. Manger acknowledging the receipt of the final pages came just as we were leaving for the airport, enroute to Athens and India. Thus the effort which began in March 1974 finally came to an end almost eight years later.

Chapter I (Mathematical Preliminaries): It was clear to me from the outset that the book had to begin with a chapter including an account of differential geometry adequate as an introduction for an *ab initio* treatment of Cartan's calculus and the tetrad and the Newman–Penrose formalisms. I was uneasy about this prospect. However, already during my convalescence from heart surgery in the fall of 1977, I had been preparing myself; and had found the treatment in Lovelock and Rund's *Tensors, Differential Forms, and Variational Principles* (John Wiley & Sons, New York, 1975) just the right level for my purposes. Nevertheless, I thought that my account should be read by one who had a greater feeling for mathematical rigor than I had. For this reason, I had already arranged with Basilis Xanthopolous that he should spend the three months (April, May and June) as my research associate during my tenure as a Regents Fellow of the Smithsonian Institution at Harvard. (I later arranged that he should also spend the months of July and August in Chicago.) As it turned out, Basilis was most helpful with Chapter I; he continued to be helpful in many other ways all during the following two years I was writing the book. This will become clear in due course.

I started to make the notes for Chapter I during March and continued writing the chapter in April and May at Harvard. The earlier sections on differential geometry and Cartan's calculus were my first attempt at writing on these matters. Basilis' criticism was very useful in eliminating my lapses from mathematical rigor.

The sections on the tetrad and the Newman–Penrose formalisms were relatively straightforward since I had already focused on them in my article for the Hawking–Israel volume. But several new developments had to be incorporated. These included the general expression of the Weyl tensor in terms of the Weyl scalars; a complete discussion of the Bianchi identities (including the Ricci terms); and the sixteen eliminant relations (involving only the spin coefficients) which follow from the thirty-six Ricci identities. Also, in the section on Petrov classification, the treatment of the type-D space-time had to be modified — the version in the Hawking–Israel volume is not entirely correct.

Chapter I was completed in May; but it was typed in final form only in July.

Chapter II (A Space-time of Sufficient Generality): This chapter was to provide the components of the Riemann tensor and Maxwell's equations for the most general space-time that would be needed in the book. By providing these basic equations at the outset, one could write out the necessary field equations for any problem, that one may wish to consider.

The metric I had in mind was a generalization (allowing for a φ-dependence) of the metric Friedman and I had considered in our joint work in 1971 for treating non-stationary axisymmetric systems. But it was essential to prove the assumption made in that work (on the basis essentially of 'counting argument') that in any 3-space, an orthogonal system of coordinates can be set up in finite neighborhoods. It was astonishing that none of the experts (including Saunders McLane) whom I consulted was even aware of the existence of such a theorem. But by constant persistence, I was able to elicit from Trautman the reference to the paper by Cotton (referred to in an exercise' in Petrov's book). The 'Cotton–Darboux' theorem was inserted only in October. The rest of the chapter had been written during the months of May and June.

The formulae given in the paper by Friedman and myself were generalized in this chapter to allow for a φ-dependence. The analysis was made tractable by the device of introducing the colon derivative.

The section on Maxwell's equations similarly generalizes the treatment given in the 1979 paper (on the Reissner–Nordström perturbations) by Basilis and myself.

Chapter III (The Schwarzschild Space-time):   In   deriving   the Schwarzschild metric, I followed Synge in starting off directly in the Kruskal frame utilizing a pair of null coordinates. (But some basic misunderstandings of Synge had to be clarified.) The standard derivation in the Schwarzschild coordinates was given as an alternative. However, for both derivations, the appropriate field equations could be written down directly by suitable specializations of the general formulae given in Chapter II.

The main part of the chapter was devoted to the geodesies. I found the extant treatments, except Darwin's, unsatisfactory. But Darwin had treated only a part of the problem. The matter of treating the entire problem, *de novo*, and providing a complete classification of all the geodesies, took a much longer time than I had anticipated. An essential novelty of the treatment, which unifies the discussion, was the introduction of imaginary eccentricity and the distinction between orbits of the first and the second kinds.

While writing this chapter, it occurred to me that it would be useful to provide illustrations of the various classes of orbits. I was fortunate in getting the assistance of Garrett Toomey (a student of

Dave Arnett's). His illustrations add considerably to the treatment of the geodesics in this chapter (and also in Chapters V and VII) in the context of the Reissner–Nordström and the Kerr space-times.

The chapter ends with a description of Schwarzschild space-time in a Newman–Penrose formalism. The chapter was written during the months of June, July and August.

Chapter IV (The Perturbations of the Schwarzschild Space-time):
'At long last,' — so it seemed at the time! — in October, I began to write on matters which had been my primary concern in my series of papers in the *Proceedings of the Royal Society*. As I stated in the Bibliographical Notes, Chapter IV brought together the results and methods scattered through my various papers to provide a coherent and unified treatment of the perturbations of the Schwarzschild black hole. Besides, the chapter included an account of the theory of Xanthopolous which led him to isolate the special integral which enabled the reduction of the three equations governing the polar perturbations to a single second-order wave equation. The chapter also included an introduction to the theory of inverse scattering to the extent necessary to understand the equality of the transmission amplitudes for the axial and the polar perturbations. In writing this account of the theory of inverse scattering, I had the benefit of correspondence with Professor Deift of New York University.

The section on the 'physical content of the theory' (Sec. 32) is new. It deals with an aspect of the subject I had ignored in my writings.

And finally, the treatment of the stability of the Schwarzschild space-time was direct and simple: it differed from the usual treatments which ignore the standard theorems of the quantum theory.

Again this chapter took longer than I had expected: it was only during the first week of January (1981) that the final typed copy was ready.

Chapter V (The Reissner–Nordström Solution): While the theory of the perturbations of the Reissner–Nordström solution had been fully worked out in the two papers published during 1979, there were other

aspects that had to be considered afresh. (But even in the theory
of perturbations, the derivation of Maxwell's equations, 'already linearized' had to be considered (Sec. 44a) — an aspect of the problem
to which I had not paid any attention in my earlier writings.)

First, there was the matter of the derivation of the Reissner–
Nordström solution. It had, of course, to parallel the derivation of
the Schwarzschild solution in Chapter III; and this was not entirely
straightforward.

Next, there was the matter of the geodesies. Again the extant
treatments were inadequate and unsatisfactory. The treatment, par-
alleling the account in Chapter III, had to be developed *ab initio*.
I was again fortunate in having Toomey provide examples of the
various critical trajectories.

Then there was the matter of relating the fluxes of gravitational
and electromagnetic energies with the functions $z_1^{(\pm)}$ and $z_2^{(\pm)}$ that
I had originally taken over the required relation from Matzner. The
derivation of the relation in Sec. 47 gave me some trouble. But the
introduction of the scattering matrix (at the suggestion of R. Sorkin)
brought an element of elegance to the entire subject.

And finally, there was the matter of the so-called instability of
the Cauchy horizon. The extant accounts were either in part wrong,
or orthogonal to the spirit of the book. The treatment in Sec. 49 was
long delayed. It was completed only in August, after several discussions with Hartle. But the nth copy for this chapter, exclusive of the
last section, was completed in February (before going to Salonika),
though the final typed copy was ready only in March.

Chapter VI (The Kerr Metric): I knew that when I came to writing
the opening two chapters on the Kerr metric, I would need personal
consultations with Basilis; and I had tentatively arranged that I
would visit him in Salonika in December 1980. But the delay in
writing the earlier chapters required a postponement of the visit.
However, I did not want to postpone it too long.

There were two principal matters which I wished to consult
with Basilis. The first was in connection with the Kerr–Schild

coordinates; and the second was with respect to the complex integral of Penrose and Walker governing geodesic null motions in type-D space-times. But the earlier sections of Chapter V dealing with the derivation of the Kerr metric, the uniqueness theorems of Carter and Robinson, and the introduction of the Kerr–Schild coordinates, had to be written before going to Salonika; and I began working on these matters, in earnest, already in December while getting the final copy of Chapter V ready.

There was no difficulty in writing the sections dealing with the derivation of the Kerr metric: in the main I could follow my paper on the topic in the *Proceedings of the Royal Society*. But the part relating to the solution of $(\mu_2 + \mu_3)$ had to be corrected: in the published paper I had obtained only one of the two equations governing $\mu_2 + \mu_3$ so that a unique solution could not be obtained.

The second equation followed from the field equation, $R_{23} = 0$. The two equations $(G_{22} - G_{33} = 0$ and $R_{23} = 0)$ made the solution of $\mu_2 + \mu_3$ determinate. (The two Equations (59) and (60) became essential for certain later developments: in the derivation of the Kerr–Newman solution and in the discussion of the static-distorted black holes, both in Chapter XI.) In addition, I had also to work out, explicitly, the non-vanishing components of the Reimann tensor and show explicitly that the Weyl scalars Ψ_0, $\Psi_{,1}$, $\Psi_{,3}$ and Ψ_4 vanish and evaluate Ψ_2. The calculations were formidable and sorely taxing.

Fortunately, I had carefully prepared, some years earlier (1976?), a complete set of notes with all the gory details in the proof of Robinson's theorem. These notes made the writing of this section fairly easy. In the same way, a set of notes Basilis had worked out for me in 1978 on his method of introducing the Kerr–Schild coordinates, helped me greatly in writing these parts of the chapter. But the part dealing with the nature of the Kerr space-time, as exemplified in the Penrose diagram, was giving me conceptual difficulties. They were unraveled only much later in May. But by February 20, I had the rest of Chapter VI in the $(n-1)$ version. And so we departed for Salonika on February 21.

Chapter VII (The Geodesies in the Kerr Space-time): My principal object in going to Salonika at this juncture was to obtain a direct proof of the complex-integral of Walker and Penrose (for null geodesic motion in type-D space-times) which will not require knowledge of the spinor formalism. And I was certain that, together with Basilis, I could devise such a proof.

I had informed Basilis, beforehand, of the principal object of my visit; and he was prepared by having read particularly the papers by Stark and Connors. At first, we did not know how to proceed; but before the end of the week we had established Theorem 1 of Sec. 60. We did not, however, have the time to think about Carter's real integral for general geodesic motion in the Kerr space-time. But I felt that a straightforward extension of the ideas of Theorem 1 would yield the required result.

On returning from Salonika, I was able to prove Theorems 2 and 3. I was particularly pleased with Theorem 3 (and so was Basilis) because it provided the necessary and sufficient conditions for the existence of a Carter-type integral for general geodesic motion for type-D space-times in terms of the spin coefficients.

Chapter VII was, in the main, devoted to the integration of the geodesic equations in the Kerr space-time. There was a massive literature on the subject, but none of it was to my taste: the treatments were haphazard, incoherent, and partial at best. On this account, I decided to develop the necessary formulae independently. But it meant spending an additional month — Chapter VII took nearly three months. Again, Toomey provided the necessary illustrations to supplement my analysis. Besides, his calculations revealed certain errors of interpretation; and they were corrected only in August.

Since Penrose's discussion of the elementary process, in terms of which he wished to illustrate the way the rotational energy of the Kerr black hole could be extracted, was based on a numerical example, I had to consider this matter analytically *de novo*. The treatment of the Penrose process in Sec. 63 was more general than those found in the literature. But the Wald and the Bardeen inequalities were derived essentially in the manner of these authors.

Chapter VIII (Electromagnetic Waves in Kerr Geometry): By early
June, I was ready to start on Chapter VIII on the electromagnetic
perturbations of the Kerr black hole. By and large, this chapter
presents my published work in a self-contained manner while providing additional details on such matters as the Starobinsky identities.
At the same time, I could present the basic problems associated with
my particular transformation of the variables which are singular in
the super-variant interval. Also, the transformation theory could be
developed for general spins once and for all.

Besides, I had to deal in depth with the various questions relating to the physical content of the analytical results derived from the
one-dimensional wave-equations. The arguments had many pitfalls;
and I felt sufficiently insecure to ask Steve Detweiler to scrutinize
this chapter with care.

The matters that concerned me most in this chapter were
(1) the direct evaluation of the energy-flow across the horizon from
the known expression for the energy-momentum tensor of the electromagnetic field; and (2) the justification of the boundary conditions
used in the super-radiant interval. These matters were considered in
Sec. 76. This chapter was completed during the third week of July.
Only two months remained before I was to meet Manger in New
York.

Chapter IX (The gravitational perturbations of the Kerr black hole):
The climax of the book was reached in this chapter devoted, almost
entirely, to my own work included in the series of four papers published in the *Proceedings of the Royal Society* — a work which took
me three and one-half years (from the summer of 1976 in Cambridge
to February 1980). The organization of the material, as such, was
not a difficult problem. For, as I wrote at the conclusion of my fourth
paper: "while the manner of the final outcome was unknown, even
at the stage of Paper III, there has been, through the course of these
papers, no circumlocution on the approach towards the complete solution: the analysis has been, almost, self-propelled." The account
could, however, be made a little more straightforward by making

the gauge assumptions, $\Psi_1 = \Psi_{33} = \Psi_2^{(1)} = 0$, right from the outset. The real task was to assemble all the formulae, some 70–80 pages of them; and write them out suitably for the final copy, making sure that errors did not creep in during the transcription. But even so, this longest chapter of the book was written in the shortest time — some five weeks.

The only section which gave some difficulty was Sec. 98 in which the gravitational fluxes at infinity and at the horizon were evaluated in terms of the linearized theory (at infinity) and of the Hartle–Hawking formula (at the horizon). For the linearized theory, I found the account in Robertson's book most suitable for the purposes on hand; and for the Hartle–Hawking formula, I preferred Carter's account (in the Hawking–Isreal volume) to the original treatment by Hartle and Hawking.

Chapter IX was completed on 4th September; and that left only fifteen days before our scheduled departure to Poland. I therefore abandoned the idea of writing Chapter X on Dirac's equation and concentrated on getting all the illustrations collated with legends; and also on revising some pages in the different chapters. And there was the matter of the Prologue. I gave up my initial idea of several pages and decided instead on a short single paragraph. There was, naturally, the last minute strain; but the material was all compiled together and on September 19, it was handed over to Manger at the Kennedy Airport in New York. And we left for our fortnight's vacation with the Trautman's with some peace of mind.

Chapter X (Spin-$\frac{1}{2}$ Particles in Kerr Geometry): Even though I had separated Dirac's equation in Kerr geometry in 1976, I was not at ease with the prospect of writing a chapter dealing with spin-$\frac{1}{2}$ particles. The principal reason for the uneasiness was that (as John Friedman had shown me at the time I separated Dirac's equation) the most direct way of writing Dirac's equation in the Newman–Penrose formalism was via the spinor formalism. I had not included the spinor formalism in Chapter I; indeed, I had not studied it in depth at any time. So, even prior to our departure to Poland, I had asked John Friedman to come to Chicago and explain to me how I could set out a simple derivation of Dirac's equation "without tears." I was convinced after the meeting that I simply could not avoid an account of the spinor formalism adequate for my purposes. I tried to read the accounts of spinor analysis in the various books; but they were unsuitable for my purposes. It then occurred to me to go back to my notes of Dirac's lectures on spinor analysis which I had taken in 1932. It was fortunate that I had written the notes with great care, they provided the basis I needed. But they had to be incorporated in a dyad formalism suitable for writing spinorial equations, in curved space-times in a Newman–Penrose formalism. Some notes of his which John Friedman had given me enabled me to develop this part of the theory *ab initio*. Indeed, I had found time to prepare a complete set of notes on the spinor formalism, adequate for my book, before we left for Poland.

During our stay in Poland, I was able to supplement my notes to include a derivation of Dirac's equation, its separation, and its transformation to the form of a one-dimensional wave-equation.

It seemed to me that Dirac's equation in Kerr geometry must manifest the phenomenon of the Klein paradox; and I discussed this matter at some length with Trautman. But on examining the question in detail on returning to Chicago, I was able to show that the phenomenon is not present in Schwarzschild geometry. Whether it is present in Kerr geometry is yet unresolved.

The digression into an investigation of the Klein paradox delayed Chapter X. But the $(n-1)$ draft of the earlier parts on spinor analysis was reviewed and checked by Friedman before we left for the U.S.S.R. on October 19. The chapter was completed on our return; and it was mailed to Manger on November 16. And that left only one more chapter to write.

Chapter XI (Other Black Hole Solutions): My original idea was to title this chapter "Problems solved and unsolved" and include in it accounts of the Kerr–Newman solution and its perturbations, the distorted black hole solutions of Geroch and Hartle, and the

Hartle–Hawking interpretation of the many black hole solutions of Majumdar–Papapetrou. But the chapter was to begin with accounts of Wald's treatment of the Kerr perturbations and the Friedman–Schutz proof of the stability of the Kerr metric for axisymmetric perturbations.

The first problem on my agenda after writing Chapter X was the derivation of the Kerr–Newman solution. I thought that there ought to be a way of deriving the solution from the Kerr solution by some simple transformation. I wrote to Basilis (before going to the U.S.S.R.) asking him whether there was such a method. His response, while seemingly positive, convinced me that there was no such transformation. The problem as it presented itself had two parts. The first was to derive the pair of Ernst's equations patterned after my derivation of the Kerr metric in Chapter VI; and the second was to extend the operation of conjugation to the Einstein–Maxwell equations. The first was a task that could, with patience, be accomplished. The second seemed intractable. Indeed, the whole prospect was not encouraging; and besides, I was tired after the Russian trip (we returned on November 4). Observing my state of lassitude and frustration, Lalitha suggested that it might be helpful if I could discuss my problems with Basilis and asked "why don't you go to Greece?". The idea that I might go had not occurred to me; but now it seemed the only way. So to Basili's surprise, I called and told him that I was coming to spend the following week (December 1–9) with him. And it turned out to be a most useful and necessary trip. But before I went to Salonika, I managed (with the skin of my teeth) to derive the pair of Ernst equations by an extension of the methods used in Chapter VI.

The agenda for the week in Salonika was the following: to resolve the operation of conjugation for the Einstein–Maxwell equations governing stationary axisymmetric space-times; to obtain a derivation of the distorted black hole solutions consistent with my way of looking at these problems, and, if possible, to extend the considerations to include the explicit construction of toroidal black hole solutions; and finally, to get a clear understanding of Wald's procedure. The resolution of the first problems was more difficult than I had thought. It took us three days of very hard work. The problem of toroidal black holes appeared tractable; and we spent considerable effort on it. By week's end we thought that the problem had been solved in its essentials. But we did not make any progress in getting to grips with Wald's method.

On returning to Chicago, I first concentrated on completing the derivation of the Kerr–Newman solution. It was not as straightforward as I had supposed.

The matter of the distorted black holes was more complex than we had thought. Even for the distorted black holes with spherical topology, I had overlooked the question of their equilibrium, and the requirement of local flatness of the metric on the axis. I was able to resolve these questions after some discussions with Hartle. Because of this and other problems, I abandoned the idea of including a treatment of toroidal black holes. (Subsequently Basilis resolved the various questions; and I recommended that he publish the results on his own.)

The section on the Majumdar–Papapetrou solution had again two parts: its derivation and its interpretation. The derivation could be accomplished without difficulty, thanks to the general formulae of Chapter II (allowing for the φ-dependence). And I was helped in the interpretation by Hartle.

What now remained was the last section, Sec. 114, on the Friedman–Schutz theorem on the stability of the Kerr solution for axisymmetric perturbations. By now it was January and only ten days were left before the 'ultimate deadline' of January 10 upon which Manger and I had agreed.

I had already discussed with Friedman about his theorem with Schutz on two earlier occasions: once in November (after our return from the U.S.S.R), when I went to visit him in Milwaukee; and again in December (after my return from Salonika) when John came to visit me. On the latter occasion, John spent some three hours setting out in some detail the outlines of the proof of his theorem; and I thought at the time that I should be able to give a reasonable account of the theorem in some six or seven pages. But when at last I began to think earnestly of the problem, I discovered that the matter was not that simple. I had not fully realized that what was really required was a generalization of the variational principle (that John and I had established in our joint work of ten years ago) to allow for a non-diagonal term (g_{23}) in the metric. I did not see that I had any choice but to derive the entire generalization *ab initio* which meant that I had to do, within three weeks, what John and I had done together in some ten months. I decided that I would postpone Sec. 114 and concentrate on getting the rest of the book, exclusive of this section, ready.

During the next few days, I concentrated on getting Chapter XI, exclusive of the last section, in its final form — the remaining illustrations, the Appendix and the Epilogue.

First some remarks about the Appendix. In my papers on the gravitational perturbations, I had found the need to verify my identities numerically to assure myself that no errors in the reductions had inadvertently crept in. I concluded then that the book must include a set of tables of the various Teukolsky and other functions which play a central role in the theory. Steve, who had carried out some integrations earlier for my purposes, had recomputed several of these and completed the integrations to provide a representative sample. These tables had been typed during December; and now I wrote the necessary introductory material.

The Epilogue, which had been one of my concerns from the outset, was a matter of constant debate within myself. I decided finally on a short paragraph.

On January 7, I mailed all of the book except for the last Sec. 114. And I embarked on the long calculations needed for a really satisfactory base for the Friedman–Schutz theorem. Rarely have I felt more despondent than I was during the three weeks that followed. The work needed absolute concentration: the final outcome required that not a single error be made: it was altogether a most taxing three weeks. But all is well that ends well; and it did end well: I was able to write the last Sec. 114 (the longest single section in the book!) during the third week of January; and the final manuscript was sent to Manger on the morning of Tuesday, January 26.

On Friday, January 29, we left for Athens and India; and just as we were departing for the airport, a call from Manger acknowledged the receipt of the last section. Thus, the effort of eight years came to an end!

March 23, 1982

Postscript: Basilis was our guest in Athens and we spent Sunday (January 31) at the Acropolis.

Principal Dates

First suggestion of writing a book on 'black holes' in a letter to Denys Wilkinson on October 14, 1975. A detailed outline of the book sent on October 21, 1975. Contract signed December 1, 1975.

Inquiries concerning 'progress' of book:

March 24, 1977
November 1, 1977
October 10, 1978
October 29, 1979

The last of the Proceedings of the Royal Society papers communicated on February 19, 1980.

Began writing book on March 1, 1980.

Progress reports sent on:

July 14, 1980
October 27, 1980
March 23, 1981
May 1, 1981

Chapters I–IX, inclusive of illustrations, handed to Mr. Manger on September 19, 1981 (prior to our departure to Poland) at Kennedy Airport.

Chapter X sent on November 16, 1981, after returning from the U.S.S.R.

Chapter XI (exclusive of Sec. 114), the Appendix, and Epilogue sent on January 7, 1982.

Chapter XI, Sec. 114, sent on January 26, 1982.

Call from Manger acknowledging receipt of Sec. 114 (just as we were to leave for O'Hare enroute to Athens and India) January 29, 1982.

Began work March 1974

1. Schwarzschild perturbations	7th October 74
2. Quasi-normal modes — Schwarzschild	6th December 74
3. Kerr perturbations — axisymmetric	3rd February 75
4. Transformations of Teukolsky equations; electromagnetic perturbations	11th July 75
5. Maxwell's equations in Kerr geometry	18th September 75
6. Kerr perturbations: general	5th January 76
7. Dirac's equations in Kerr geometry	21st April 76
8. Neutrino: reflexion and transmission	2nd June 76
9. Kerr metric (derivation)	18th April 77
10. Gravitational perturbations I	2nd May 77
11. Gravitational perturbations II	20th June 77
12. Gravitational perturbations III	22nd August 78
13. Reissner–Nordström I	28th August 78
14. Reissner–Nordström II	31st October 78
15. One-dimensional potential barriers	30th July 79
16. Gravitational perturbations IV	25th February 80
17. Mathematical theory of black holes	26th January 82
18. On crossing the Cauchy horizon	18th June 82

Mathematical Theory of Black Holes

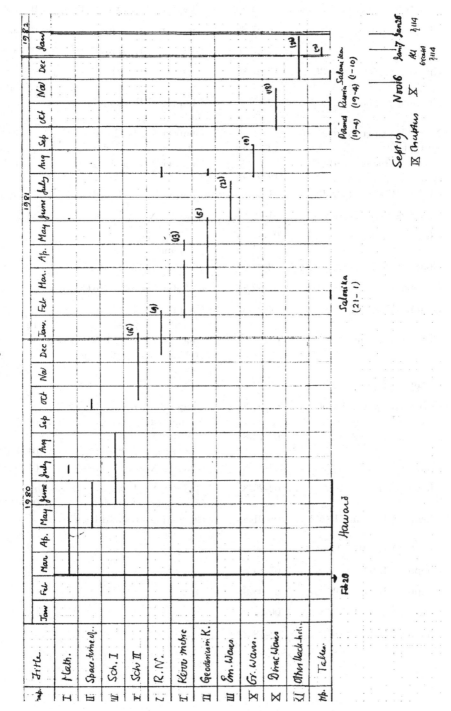

Chapter II, nth 10-13-80

Chapter IV, notes 10-27-80

Chapter IV, nth 11-07-80

Chapter IV, nth, §30–35 12-01-80

Chapter V, notes 12-15-80

Notes, Reissner–Nordström
solution 12-19-80

Notes, Reissner–Nordström
solution 12-22-80

Chapter IV, $n - 1$ draft 01-05-81

Chapters II, IV, xerox 01-16-81

Chapter VI, notes 02-02-81

Chapter VI, notes 02-04-81

Chapter V, nth 02-09-81

Chapter V, xerox 03-09-81

Chapter VI, nth 03-13-81

Chapter VII, notes 03-23-81

Corrected pp. 64, 65, 66 04-03-81

Chapter VI, 1–62 04-13-81

Chapter VII, nth 04-17-81

Chapter VII, nth, 1–50 04-29-81

Chapter VII, nth, 51–102 05-01-81

Chapter VII, 105–117 05-06-81

Chapter VI, 62–71 05-13-81

Chapter VII, pp. 1–99 06-05-81

Lemmas, pp. 100–7 of VI 06-12-81

Chapter VIII, 1–48 nth 06-18-81

Chapter VIII, nth, 37–66 06-22-81

Chapter VIII, nth, 1–86 06-30-81

Chapter VIII, xerox 1–29 07-06-81

Chapter VIII, nth, 75–80, 19a,
1–70 07-23-81

Chapter VIII, changes, 48, 50,
60 ff 08-07-81

Chapter IX, final draft
1–156 09-04-81

Chapter X, nth, 1–42 11-02-81

Chapter X, final copy, 1–4 ... 11-16-81

Chapter X, final pages 11-18-81

Acknowledgments,
Prologue etc. 12-10-81

Chapter XI, notes, 25–66 12-14-81

Chapter XI, notes (to Crete) . 12-16-81

Chapter XI, 1–12 notes, 1–26
formulae (expres) 12-21-81

Chapter XI, nth, 1–50 (xpres)
and repeated mailing of
Dec. 16 12-28-81

Chapter XI, nth, §113, 51–66
+ final copy, 1–42 01-04-82

Chapter XI, nth, §113, 51–66,
+final copy, 1–42 01-04-82

Chapter XI, final, 42–54 01-06-82

Chapter XI, 1–46 and R1–R6 plus
Final figures from R1–R6 01-18-82
(field equations and their
linearization)

Outline as Originally submitted on October 21, 1975
The Schwarzschild and the Kerr Black Holes

Outline as Revised on October 16, 1979
The Relativistic Theory of Black Holes

I.   INTRODUCTION

II. MATHEMATICAL PRELIMINARIES

   1. Exterior Calculus

   2. The Cartan Calculus

   3. The Tetrad Formalism

   4. The Newman–Penrose Formalism

III. THE FIELD-EQUATIONS FOR A NON-STATIONARY
   AXISYMMETRIC SPACE-TIME

   1. Einstein's Equations

   2. Maxwell's Equations

   3. Einstein–Maxwell Equations

   4. Generalization to Allow First-Order Non-Axisymmetric Perturbations

IV. THE SCHWARZSCHILD METRIC

   1. Derivation by Cartan's Calculus

   2. Synge's Derivation

   3. The Kruskal Frame

   4. The Geodesies in the Schwarzschild Metric

      i. Null Geodesies

      ii. Time-Like Geodesies

V.   THE PERTURBATIONS OF THE SCHWARZSCHILD METRIC

   1. The Metric Perturbations

   2. The Equations Derived from the Newman–Penrose Formalism

   3. The Relationship Between $z^{(+)}$, $z^{(-)}$ and y

   4. The Dual Transformations

   5. The Reflexion and the Transmission Coefficients

      i. Their Equality

      ii. The Infinite Hierarchy of Integral Equalities

   6. Quasi-Normal Modes

   7. Concluding Remarks

CONTENTS

(b) The representation of the Weyl, "the Ricci, and the Riemann tensors"

(c) The commutation relations and the structure constants

(d) The Ricci identities and the eliminant relations

(e) The Bianchi identities

(f) Maxwell's equations

(g) Tetrad transformations

9. The optical scalars, the Petrov classification, and the Goldberg–Sachs theorem

(a) The optical scalars

(b) The Petrov classification

(c) The Goldberg-Sachs theorem

Bibliographical Notes

II. A SPACE-TIME OF SUFFICIENT GENERALITY

10. Introduction

11. Stationary axisymmetric space-times and the dragging of inertial frames

(a) The dragging of the inertial frame

12. A space-time of requisite generality

13. Equations of structure and the components of the Riemann tensor

14. The tetrad frame and the rotation coefficients

15. Maxwell's equations

Bibliographical Notes

III. THE SCHWARZSCHILD SPACE-TIME

16. Introduction

17. The Schwarzschild metric

(a) The solution of the equations

(b) The Kruskal frame

(c) The transition to the Schwarzschild coordinates

18. An alternative derivation of the Schwarzschild metric

19. The geodesics in the Schwarzschild space-time: the time-like geodesics

(a) The radial geodesics

(b) The bound orbits ($E^2 < 1$)

(i) Orbits of the first kind

January 26, 1982

POSTSCRIPT: 1982, A YEAR THAT PASSED

In a week it will be fully a year since I finished my book. A year has passed. How?

First, the month of February was spent in India. A week in Bombay with a day's trip to Ajanta and Elora (including a lecture at Aurangabad) and a visit to the Tata Institute; four days in Ahmedabad and three lectures; three days in Delhi including a meeting with Indira Gandhi and a lecture at Delhi University; two days in Darjeeling, a view of the Himalayan range, and a lecture at Bagdogra; three days in Madras (including a lecture); and ten days in Bangalore (two lectures at the Raman Institute and one at the Indian Institute of Science). I should have enjoyed my stay in India, if I did not have to give so many lectures and spent the time, instead, quietly at Savithri's and Balakrishnan's. I wish that we had gone to India on our own without the sponsorship of the C.S.I.R. I was already very tired; and all the public functions did me no good.

We returned to Chicago on March 4. The first 'job' I had to attend was to revise my Vikram Sarabhai lectures which Bimla had transcribed on tapes. Soon after, the galleys of my book began to arrive; the last of them were returned on June 23. Meantime, I wrote the paper with Hartle on "The crossing of the Cauchy horizon of a

Reissner–Nordström black hole"; it was sent to the Royal Society on June 14. Most of July and August were devoted to assembling the material for my Eddington Centenary lectures and preparing the nth copy. During the first three weeks of September we had a holiday in Turkey: Istanbul, Ederni, Anatolia and the battlefield at Gallipoli. Returning from Turkey, I had to get the manuscript of my Eddington lectures in its final form. In Cambridge for the lectures during October 18–25. (The lectures were published as a small book by the Cambridge University Press.) The page proofs were waiting on our return; and they and along with the Index took up all of November. Then I had to write up my paper for the 1981 Byurakan Symposium on Principles of Invariance. In some ways, I was glad for this opportunity to write the evolution of my work on radiative transfer during 1943–48. When that was written and sent, the proofs of the Index came: and that was sent off on December 30. On returning the proofs of the Index, the last of my duties towards the book came to an end. And we decided to keep a long postponed promise to Cesco that we would visit him in San Juan. It also gave us the occasion to spend ten restful days in Bariloche. And that was how the year passed.

January 22, 1983

THE BEGINNING OF THE END
(1983–1985)

In 'The Year that Passed' written in January 1983, I said that that was probably the last of these installments. In so writing, it was my expectation (and, partially, my intention) that I would give up serious practicing of science. That intention was not adhered to, as the list of my publications since January 1983 will show. While four of them are the printed versions of lectures, the remaining seven represent as serious a scientific effort as any that I have undertaken. Besides, the two and half years that have elapsed have been full of distractions as is evident from the brief chronology. I shall begin from where I left off.

On returning from our vacation in Argentina, for some reason I cannot quite recall I began to wonder if the one-dimensional Schrödinger equation, for the particular form of the potentials that occur for the Schwarzschild and the Reissner–Nordström perturbations, will allow some explicit solutions. It was immediately clear that such solutions exist for frequencies which make the Starobinsky constant vanish. I first thought mistakenly that these special solutions belonged to purely damped quasi-normal modes. But Wald pointed out my error and further that the condition — the vanishing of the

Starobinsky constant — led to algebraically special perturbations belonging to purely ingoing or purely outgoing monochromatic waves. Once having realized this, I formulated the problem precisely and was able to obtain the explicit form for the 'wave functions' belonging to these special perturbations. The formulation of the problem made it clear that these algebraically special perturbations had a bearing on the alternative factorization of the Starobinsky constant provided by the transformation theory. The consistency of the two alternative ways of specifying these special perturbations required the identity of two polynomials of degree nine. I derived the identity before we left for Japan, realizing that its verification would be long and arduous. I completed the verifications, a power at a time, during the odd spare times I had during our stay in Japan: in the mornings or in between lectures. I completed the paper soon after our return from Japan. One interesting fact which emerged — not known or expected before — was that there exist special perturbations of the Reissner–Nordström black-hole which are characterized by purely ingoing and purely outgoing determinate mixtures of gravitational and electromagnetic waves.

In Japan, I gave many lectures — popular and otherwise. I did not learn anything myself and I doubt if my lectures benefited anyone either. But the drive over the mountains to Tokyo from Nagoya with the Hayakawas and later the visit to Hiroshima and the Peace Museum were memorable.

During spring, following our return from Japan, I gave a course on cosmology — a repetition of one I had given two years earlier. But the attendance was so poor that I discontinued half-way during the course. Also, I was occupied with the opening address I was to give at GR-10 in Padua in early July.

Returning from Padua and Rome, I spent some time on the Bethe lectures I was to give in October. The summer was uneventful and I was not sure what I wanted to do. But the calm was to be shattered soon after our return from Cornell with the announcement from Stockholm. Apart from the disruption caused by innumerable telegrams and letters (and by the necessity of rewriting the articles that were being written for *Physics Today* and elsewhere), I had to concentrate on the lecture I was to give in Stockholm. In preparing the lecture, I realized that the criterion for the onset of relativistic instability had to be investigated afresh for stars with high degrees of central condensation. The problem was to find how the constant K in the inequality, $R \leq \frac{2GM}{C^2} \frac{K}{\Gamma_1 - 4/3}$, behaved as the polytropic index tended to 5. Since I was pressed for time, I consulted Norman Lebovitz and we soon found that it could be obtained from Emden's formula for the potential energy Ω for a polytrope. Eventually, we wrote a short paper for the *Monthly Notices*.

The visit to Stockholm was not the only interruption. In January, we had to go to Zürich for the Pauli Lectures. By mid-January, I was able to return to the problem of colliding waves in general relativity. But I must go back to the meeting in Padua and earlier.

When the Letter by Khan and Penrose on the problem of colliding waves appeared in *Nature* in 1971, I was instantly intrigued. I remember asking Andrzej Trautman (who was then spending six months in Chicago) to explain to me the contents of this Letter. But I don't think that I understood the full meaning, though Penrose's statement that one may have here another example of a generic singularity in general relativity remained as a matter for further study. Other things intervened and my curiosity was pushed aside. My curiosity was roused once again in 1978 when Nutku pointed out to me in a letter that his generalization of the Khan–Penrose solution to allow for non-parallel polarizations of the colliding waves involved a 'complexified version' of my simplest solution of the X and Y equations included in my paper on the derivation of the Kerr metric. I decided that I must explore this matter on *ab initio*. But I had to postpone looking into the matter since at that time I was too deeply involved with black holes. Finally, during the summer of 1983 (after I had completed my paper on algebraically special perturbations), I recalled Nutku's observation of 1978 and wondered in a vague sort of way. An accidental circumstance changed my interest from a casual one to a more active one. It happened this way.

In July, I had given the opening address at GR-10 on "The mathematical theory of black holes." After the lecture, Valeria Ferrari (whose thesis Ruffini had sent me at an earlier time) came and talked to me; and she seemed quite knowledgeable about my work. Later, when we were in Rome for a 'workshop' arranged by Ruffini, and I had asked to be excused from giving any lecture, Ferrari suggested that instead of my giving a formal lecture, there could be an informal meeting at which various interested persons could ask me questions; and I agreed. She conducted the meeting, magisterially directing the discussion along various lines: she seemed quite aware of my work in detail. It is possible that at the end of the meeting, I suggested to her that perhaps she might consider spending some time with me at Chicago.

Returning to Chicago, I forgot all about it although Ruffini called me once or twice about the possibility of Ferrari's visiting me at Chicago. In any event, a week after the Stockholm announcement, Ferrari appeared at my office. And since she had come, principally on her own to work with me, I felt obliged to suggest a problem to her. I asked her to go through my analysis of stationary axisymmetric systems in my book and transcribe the work appropriately for a space-time with two space-like Killing-vectors. She did this quite well. And she showed that an Ernst equation emerged in place of the X and Y equations; and that the simplest solution of this Ernst equation directly yielded the Nutku-Halil solution. After some four weeks of hard work she returned to Rome. I left the matter aside since the visits to Stockholm and Zürich were pending.

In spite of all the distractions during the fall, I did think at odd times how one should approach the problem of colliding waves in the Einstein–Maxwell framework. I took the occasion of Penrose's visit during the first week of December to spend a few hours with him on Sunday, 4th December, discussing this problem. Penrose thought the problem would be a difficult one, reiterating his concern over the impossibility of impulsive waves in electromagnetic theory and how one would avoid the occurrence of a square root of a δ-function.

Soon after my return from Zürich, I went over the entire analysis that Ferrari had left with me; and I went further in evaluating the spin-coefficients and the Weyl and the Ricci scalars. But I still did not understand too well the nature of the singularities at the null boundaries. (All of these became clear much later when working with Xanthopoulos on the same problem in the context of the Einstein–Maxwell theory.) But to complete the paper with Ferrari, I went to Rome in March. The joint paper was completed during the week I was in Rome. It was sent for publication soon after my return. (The formulae specifying the character of singularities on null boundaries were corrected in proofs.)

Already before I went to Rome I had transcribed the basic equations for the Einstein–Maxwell case. But no coupled Ernst equations emerged at the stage of χ and ω; they emerged only at the stage of the potentials, Ψ and Φ. It became clear that what one required was a solution of this latter equation which in the limit $Q_* = 0$ will reduce to the solution $Z = \chi + iq_2$ appropriate to the vacuum. I had realized this fact soon after my return from Rome. But I left the matter aside since my time was taken up with my course on the Theory of Solitons that I was then giving. Also, since Xanthopoulos was to come in June, I decided that we would work on the problem together during the summer and fall.

Soon after Xanthopoulos arrived in early June, he proved the basic Lemma that enabled one to write the solution for the Ernst equation appropriate to the Einstein–Maxwell equations which, for $Q_* = 0$, will reduce to the Nutku–Halil solution. The algorithm for obtaining the complete solution was clear. Basilis did most of the work at this stage, since I had to work on my revision of the last section of my *Ellipsoidal Figures of Equilibrium* for the Dover edition.

The basic solutions for χ_e and q_{2e} were obtained by early August, I provided the simpler and the more direct manner of derivation. The solution for $\nu + \mu_3$ was essentially guessed by Xanthopoulos; but it was not difficult to derive it systematically from the governing equations once one knew what to look for.

The problems relating to the verification of the jump conditions and the nature of the singularities along the null boundaries gave rise to conceptual difficulties; but they were eventually solved. The final paper was written during the latter part of September; and we managed to send the paper to the Royal Society by the last week of September at which time Xanthopoulos had to return to Greece and I had to go to Syracuse.

But already during September I began to think about the problem of colliding impulsive waves coupled with fluid motions. I very soon realized that the case, $\epsilon = p$, could be solved by essentially the same methods as in the papers with Ferrari and Xanthopoulos. I showed my preliminary calculations to Xanthopoulos and asked him to check the hydrodynamic equations. He went further and showed that the basic hyperbolic equation for the stream function Φ took its simplest form in null coordinates. At this stage neither of us knew how to solve this hyperbolic equation. It also became clear that the matter of extending the solution beyond the null boundaries was not going to be straightforward. The problem had to be laid aside because of our impending visit to India.

Besides working on the joint paper I spent a considerable fraction of the summer preparing and thinking about "The pursuit of science and its motivations." During a weekend in August, which we spent with the Cronins at their summer cottage, we discussed at great length many aspects of the motivations for pursuing science.[m] Since my intention was to read a carefully prepared text, it was important that the manuscript was completed before we left for India on October 20.

The main purpose for going to India in October was to attend the Golden Jubilee Celebrations of the Indian Academy of Science and also for the formal presentation of Ramanujan's bust. But all the arrangements had to be cancelled because of Indira Gandhi's

---

[m]I should remark that in this discussion as in the earlier discussions over the years, between ourselves and with others. Lalitha was always an active and a stimulating participant.

assassination. We nevertheless stayed on in India, since we had a commitment to be in London on November 30.

Penrose visited us in London. We talked about various problems over tea. And in the evening we went to a performance of *The Marriage of Figaro* at Covent Garden.

On returning from India in early December, I thought a good deal about methods of solving the hyperbolic equation for Φ. In the first instance, I naturally wanted to determine whether one could find an explicit formula for the Riemann function. I had written to Fritz John at the Courant Institute already before leaving for India, inquiring about the catalogue of hyperbolic equations for which the Riemann function was known — a catalogue to whose existence he refers in one of his books. In December, I received a letter from John saying that he could not trace any such catalogue but referred me to Cathleen Morawetz. I had some interesting correspondence with her; but it turned out not to be relevant for the problem on hand. Also, to my surprise I found that the equation for Φ was one of the standard equations in Copson's book on partial differential equations (which I had borrowed from Narasimhan). The discovery of the Riemann functions in Copson's book, however, led to a wrong trail: the kind of solution which I had thought was relevant, namely, that $\Phi = $ constant on the null boundaries, led to a triviality; and the trail had to be abandoned.

I next concentrated on finding whether the required solution for Φ can be found by a separation of the variables. I was able to find the exact separable solution which guarantees the positive-definiteness of ϵ in Region I. The matter of extending the solution beyond the null boundaries now became the principal problem. I discussed the problems of the extension with Bob Geroch. He suggested that Regions II and III could be made flat by *fait accompli* and that such a solution would be acceptable if one was led to singularities no worse than δ-functions. I could soon show that this was the case. And again the matter had to be put aside because of our return visit to India in February.

On returning from India I found that Curt Cutler, one of Geroch's students, had shown that the discontinuity along the null boundaries could be eliminated by a coordinate transformation. Everything now seemed to have fallen in place. Indeed, I wrote an $(n-1)$-draft of the paper along these lines. But at this stage a discussion with Lee Lindblom and Cutler showed that the solution violated causality and therefore had to be abandoned.

Left in this predicament, it occurred to me that perhaps the solution could be extended into Regions II, III and IV by finding solutions in other gauges and coordinates which did not require one to be restricted to Region I. There were indeed two other gauges to explore. Because a lot of detailed calculations were involved, I decided to go to Crete so that Xanthopoulos and I could get the entire work completed in a week of effort. We were able to accomplish this task (during the week March 8–17). While we found two extremely interesting classes of solutions, we did not resolve the basic problem. (The new solutions were later the subject of a second paper we wrote in June.) And it occurred to me quite suddenly that perhaps it may be useful for me to go to Houston to discuss the problem directly with Roger Penrose. The decision had to be made instantly because that was his last week in the United States; and so I called Penrose and decided to go as soon as it was convenient for him. The meeting in Houston proved to be decisive. After I had explained the problem and the nature of the impasse which we thought we had reached, Roger made the astonishing suggestion that perhaps Regions II and III were filled with null dust. I could not verify his suggestion straightaway, since I did not have all the necessary notes and calculations with me.

The following day, on returning from Houston, I was able to verify all the necessary requirements for null dust prevailing in Regions II and III. Penrose's suggestion was vindicated. I called to tell him of the successful completion of the project.

It should be recorded, however, that already in September, Basilis had formally noticed the possibility of null dust in Regions

II and III; but we ignored it as manifestly untenable and we did not check any of the necessary requirements. It was riot difficult now to complete the paper. But the surprising denouement gave rise to much discussion among Geroch, Lindblom and the students. I was not involved in these discussions. But periodically Cutler gave me some written summaries; and I also talked to Lindblom. And while talking about these matters with Lindblom, he suggested that I might examine the problem with the assumption that null dust prevailed in Region I. I am afraid that in deriving the basic hydro-dynamic equations, by an oversight I was led to the wrong equations. Basilis pointed out the error. When the correction was made, the solution was very simple and straightforward. The analysis could be completed in a day; but the result that one can have two entirely different solutions in Region I with the same solutions in the rest of the space-time gave rise to considerable discussion. I had difficulty in persuading even Xanthopoulos to my point of view. Even though the paper was written and sent to the Royal Society in August, I had a continuing feeling of uneasiness.

The months of August and September were devoted to reading through the entire *Mathematical Theory of Black Holes*, checking for misprints that could be corrected in its third printing scheduled for later this year. And now I could be content in the knowledge that I had done my utmost with respect to the book.

September 25, 1985

S. Chandrasekhar

PAPERS:	Acceptance date	Communication date
On algebraically special perturbations of black holes, *Proc. R. Soc. Lond. A* **392** (1984) 1–13.	23 May 1983	16 May 1983
On the onset of relativistic instability in highly centrally condensed stars (with N. R. Lebovitz), *Mon. Not. R. Astr. Soc.* **207** (1984) 13–16.	12 Jan. 1984	30 Dec. 1983
On the Nutku–Halil solution for colliding impulsive gravitational waves (with V. Ferrari), *Proc. R. Soc. Lond. A* **396** (1984) 55–74.	27 Mar. 1984	22 Mar. 1984
On colliding waves in the Einstein-Maxwell theory (with B. Xanthopoulos), *Proc. R. Soc. Lond. A* **398** (1985) 223–259.	2 Oct. 1984	26 Sept. 1984
On the collision of impulsive gravitational waves when coupled with fluid motions (with B. Xanthopoulos), *Proc. R. Soc. Lond. A*	15 May 1985	9 May 1985
Some exact solutions of gravitational waves coupled with fluid motions (with B. Xanthopoulos), *Proc. R. Soc. Lond. A*	10 June 1985	4 June 1985
On the collision of impulsive gravitational waves when coupled with null dust (with B. Xanthopoulos), *Proc. R. Soc. Lond. A*	12 Aug. 1985	6 Aug. 1985

LECTURES

1. Marian Smoluchowski as the founder of the physics of stochastic phenomena, *Postepy Fizyki* **35** (1984) 585–595.
2. On stars, their evolution and their stability, in *Les Prix Nobel en 1983* (The Nobel Foundation, 1984), pp. 55–80.
3. The mathematical theory of black holes, in *General Relativity and Gravitation*, eds. B. Bertotti *et al.* (D. Reidel Publishing Company, Dordrecht, Holland, 1984), pp. 5–26.
4. The general theory of relativity: Why "it is probably the most beautiful of all existing theories", *J. Astrophys. Astron.* **5** (1984) 3–11.
5. The pursuit of science: its motivations, *Current Science* **54** (1985) 161–169.

TRIPS

1983:

14 March–17 April, Japan (Kyoto, Nagoya, Tokyo, Osaka, Hiroshima; return to Tokyo)

22 April, Case Western Reserve University, Cleveland, Michelson-Morley Award

3–19 July, GR-10, Padua, Rome. Opening Address

2–14 October, Cornell University, Ithaca, New York. Bethe Lecture (two public lectures, three colloquia)

19 October, Nobel Laureate announcement

5–15 December, Stockholm, Sweden: Royal Swedish Academy, Nobel Award Lecture

1984:

8–15 January, Zürich, Switzerland: Pauli Lectures (Jan. 9, 10, 12); Dr. Tomalla Prize

5–12 March, Rome, discussions with Valeria Ferrari

30–31 May, D. Sc., Queens College of the City University of New York, Flushing, N.Y.

4–6 October, Syracuse University, Syracuse, New York, lecture

20 October–27 November, India. Oct. 26, Birla Award Nov. 22, Bose Lecture

28 November–1 December, London, Royal Society, Copley Medal

1985:

2–19 February, India. 6 February, Golden Jubilee Lecture

8–17 March, Crete, discussions with Basilis Xanthopoulos

17 April, Houston, Texas, discussions with Roger Penrose

30 April, Princeton University, Physics Department, Princeton, New Jersey. Hamilton Lecture

26 June–6 July, Canary Islands (Tenerife) and Madrid, Spain, Inauguration of the Institute de Astrofisica de Canarias, Tenerife

12–14 June, Washington, D. C., Dinner at White House during State Visit of Rajiv Gandhi

26 June–6 July, Canary Islands (Tenerife) and Madrid, Spain, Inauguration of the Instituto de Astrofisica de Canarias, Tenerife

15–26 August, Spring Green, Wisconsin, vacation

27 October–9 November, Iraklion, Crete, informal discussions with Basilis Xanthopoulos

19–20 November, University of Florida, Gainesville, Florida, informal discussions with relativists including Steven Detweiler

Continued Efforts I
(September 1985–May 1987)

During my visit to Crete in March (1985), I was principally occupied with the new solutions in the other two gauges and coordinates (besides the ones appropriate for Region I for colliding plane-fronted waves). While I was so occupied, Basilis occupied himself with working out the solution (on a computer) that follows from the simplest solution $p\eta + iq\mu$ of the Ernst equation for derived $E^\dagger$ for $\Psi + i\Phi$. I was not then very much interested in following this trail: my interest was diminished by Basilis's having to use a computer even to obtain the solution for $\nu + \mu_3$. But Basilis was quite insistent that the paper should be a joint one: and in July or August, I received from him his calculations as well as a preliminary draft of the paper. I put these aside since I was too involved, first with the paper on the null-dust and then in my scrutiny of my *Mathematical Theory of Black Holes* for misprints for correction in its third printing. I was able to examine what Basilis had done only by mid-September.

Since the paper was to be a joint one, I naturally wanted to scrutinize Basilis's manner of derivation. I soon found that the solution for $\nu + \mu_3$, far from requiring a computer, could in fact be obtained very simply from the known solution for the Nutku–Halil metric.

And I was also puzzled by the criterion for the choice of the constant of integration in the solution for q_2. But I did not do anything about the long and complicated formulae that Basilis had obtained for the Weyl scalars since, at this stage, they were all believed to diverge (as expected!) on the arc $u^2 + v^2 = 1$.

By early October, I had an $(n-1)$ draft of the paper. Since I was extremely tired at this time, Lalitha and I decided that we would go to Crete for two weeks (Oct. 26–Nov. 10) partly as a vacation and partly to "wind up" the loose ends of my collaboration with Basilis.

We were fortunate in the weather we had: we spent a large part of our time sitting by the seaside and reading. Basilis took us out for day-long drives in the beautiful mountainous countryside; and he checked the $(n - 1)$ draft of the paper.

On returning from Crete, I wrote the n version of the paper. While the paper was being typed, Basilis called to say that he had found some errors in his evaluation of the Weyl scalars; that when they were corrected, none of them diverged on $u^2 + v^2 = 1$; and that in particular, Ψ_2 had the extremely simple form:

$$\Psi_2 = \frac{1}{2}(1 - p\cos\psi - iq\cos\theta)^{-3}.$$

The typing of the paper had to be abandoned: there was something quite unexpected to explore.

My first reaction was: if the expression for Ψ_2 is that simple, why need a computer to evaluate it? There must be a simple way of deriving it *ab initio*. Clearly what was needed was to rewrite the expressions for the Weyl scalars, given in terms of the Ernst function

$\xi = (\chi + iq_2 + 1)/(\chi + iq_2 - 1)$, in terms of $E = (\Psi + i\Phi + 1)/(\Psi + i\Phi - 1)$.

When this was done, I found that the simple expression for Ψ_2 followed at once. It affected the corresponding reductions for Ψ_0 and Ψ_4. Since I wanted the expressions to be checked, I put them aside for Basilis to do the checking after his arrival in January. And I turned my thoughts to how the space-time was to be viewed when no curvature singularity developed in $u^2 + v^2 = 1$. I discussed the matter with Wald and Geroch; but to no avail. Again, I put these thoughts aside and returned to a problem I had thought of while we were in Crete in November: The problem of cylindrical waves and space-times with two space-like Killing vectors, the orbits of one of which are closed.

It was straightforward enough to work out the formal theory and derive the appropriate cylindrical Ernst equation. The possibility of separable solutions describing monochromatic waves almost stared at my face. It was possible to complete the solution and determine its salient properties. The radial function satisfied a nonlinear Bessel equation. It occurred to me that this equation might have occurred in the theory of solitons. So I wrote to Ablowitz inquiring whether he was familiar with the equation. He confirmed that the equation was indeed familiar to him as a special case of Painleve transcendent III. I had Chris Habisohn (a student of Geroch) integrate the equation (on a computer) and obtain some sample solutions. Also, I could easily generalize the theory to include coupling with electromagnetic fields and an $(\varepsilon = p)$-perfect third. Two problems however remained: to understand the meaning of the "C-energy" and to obtain the asymptotic behavior of the radial functions. The correct asymptotic behavior was derived by Persides after I had suggested its form based on some comments of Ablowitz. But the real meaning of the C-energy escaped me at this time; and neither Geroch nor Wald could enlighten me to my satisfaction. Except for these two items, I had worked out the full theory when Basilis arrived in mid-January. And I put the cylindrical waves aside and turned my thoughts once again to the "new solution."

I showed him what I had done: and in particular, my reduction of the formulae for the Weyl scalars. It soon became evident that the extension of the space-time beyond $u^2 + v^2 = 1$ was the primary problem. I had no idea as to how one might accomplish it: and neither did Basilis or Geroch. It occurred to me (as it had in April of 1985) that I should go to Oxford and discuss the matter with Roger Penrose. So I called him and asked him if I could come to Oxford to talk to him about the problem. He was surprised at my wanting to undertake the trip to England just to see him!

I left for Oxford on Sunday, January 26. I had some two or three hours of discussion on Monday evening (allowing time for dinner at Wadham). And I had an hour on Tuesday morning. By this time Penrose had clearly understood the nature of my problem. Later that evening, after dinner, we again talked about the matter of the extension quite specifically. And when we parted at about 11 p.m., Penrose had made a very specific suggestion as to the kind of transformation that might provide the required extension. Early next morning, I left by bus for Heathrow. And on the plane returning to Chicago, I was able to carry through successfully Penrose's suggestion. On reaching home later that evening, I called Roger to tell him that his idea had worked and told him also that I was in total amazement of his astonishing insight.

While on the plane, I had carried out the extension only for the asymptotic form of the metric near $u^2 + v^2 = 1$. The extension of the exact metric gave no difficulty; and this phase of the problem was completed within a day or two after my return from Oxford.

During our discussion in Oxford, Penrose wondered whether the solution we had was of type-D. I was not sure at that time. But Basilis was able to show that it was type-D in Region I. The space-time was indeed isometric to the Kerr space-time. But the precise nature of the geometry was not clear immediately. It was soon resolved. But it took another week or two to clarify the entire problem and realize that the extended space-time had time-like hyperbolic arc singularities.

At this stage two problems still remained: the separation of the Hamilton–Jacobi equation and the maximal analytic extension of the space-time. The former was straightforward; but the latter took considerable discussion. A number of related matters had to be understood: the disposition of the null cones along the null boundaries of the extended space-time; and the best manner of exhibiting the nature of the space-time. However, all these matters were mostly clarified when I left for Geneva (for a colloquium at Cern) and München (for an invited paper at an ESO conference arranged by

Lo Woltjer) on March 12. The constancy of surface gravity on the null horizon was proved much later in the proof stage.

Both at Cern and at Münich, I spoke principally about colliding waves, concentrating principally on the transformation of null dust into an $(\varepsilon = p)$-perfect fluid. Also at Münich, I had a number of discussions with Oscar Reula (a former student of Geroch) concerning the C-energy. He pointed out that the Ernst equation followed from a Lagrangian density; and that the associated Hamiltonian density led directly to the C-energy. At last, I understood what the C-energy really signified.

On returning from Münich, I wrote the $(n-1)$th and then the nth version of the paper with Basilis. The completed paper was sent in to the Royal Society on April 7. Next, I started on the paper on cylindrical waves: and this too was completed before I went to Boston on May 1 for my lecture at Northeastern University.

In many ways April was a cruel month: Lalitha's surgery for cataract on April 1, trips to Detroit (on April 21) and Boston (on May 1), and the completion of the two papers. And after all that, I still had to write my foreword for Eddington's *Internal Constitution of the Stars*.

With the completion of the two papers — on cylindrical waves and on the new type of singularity — I felt that I had come to the end of my efforts on the theory of colliding waves and related problems; and that I should begin to think seriously about the Karl Schwarzschild lecture that I was to give in September. The lecture was to be a major effort on my part — but more of this later. But a chance 'discovery' during the Memorial Day weekend resulted in further efforts and the hectic pace of the preceding months was to continue for several more months. It came about this way.

During Nutku's visit in February — I had arranged for his visit — he had referred a good many times to the importance of finding a two-parameter generalization of the Bell–Szekeres solution, and in particular, of finding a solution with two or more parameters of an Ernst equation in toroidal coordinates that he had derived. At that time, I did not know anything about the Bell–Szekeres solution — indeed, I was inclined to be contemptuous of it — but I was intrigued to find out more about the Ernst equation in toroidal coordinates. Actually, Nutku was never explicit about the nature of his problem: he tended to be 'secretive' about it — at least, so I thought. I therefore wrote to Nutku asking him for some specific information about his Ernst equation. Meantime, I tried to find out more about the Bell–Szekeres solution and how it was a solution of the Einstein–Maxwell equations as Xanthopoulos and I had written them out in our joint paper. I asked Basilis to explain it to me since he seemed knowledgeable about the matter. It transpired that the Bell–Szekeres solution followed from the special solution, $Z = 1$ and $H = \eta$ of the coupled equations governing Z and H. It was straightforward to find the complete generalization of this solution in the hyper-surface orthogonal case: but Basilis did not think that there was anything new to be learned from my general solution (though he changed his mind later). I continued to think about the problem; and the 'chance discovery' to which I referred earlier was the simple realization that for $Z = 1$, H satisfied the Ernst equation for a vacuum with all the consequences that it implied.

Realizing this, I concluded (too hastily as it turned out) that $E = p\eta + iq\mu$ must provide the two-parameter generalization that Nutku was seeking. During the same Memorial Day weekend, I completed the solution for this case and confirmed that the space-time in which the electromagnetic wave was propagated was conformally flat with the gravitational field confined exclusively to the impulsive waves. I showed the solution to Xanthopoulos the following day. He seemed skeptical that the solution could really be different from the Bell–Szekeres solution; and later in the day he proved conclusively that the solution I had derived with the two parameters p and q could be reduced to the Bell–Szekeres solution by replacing x^1 and x^2 by a suitable linear combination of them with constant coefficients.

In spite of my initial failure in going beyond the Bell–Szekeres solution, I was convinced that the reduction of the Einstein–Maxwell equations to the vacuum Ernst-equation (albeit, in the special case $Z = 1$) had the potential to lead to further solutions of interest. I had, for example, already obtained (during the same weekend) the stationary, monochromatic-wave solution in cylindrical geometry; and had decided to add a new Appendix to my paper on cylindrical waves then in press. A specially interesting feature of this last solution is that a time-independent space-time supports monochromatic electromagnetic waves.

Since the solution $E = p\eta + iq\mu$ led only to a non-degenerate form of the Bell–Szekeres solution, it was clear that one way to go beyond it was to consider its Ehlers transform, $\tilde{E}$, and obtain a new one-parametric family of solutions. The consideration of the Ehlers transform, $\tilde{E}$, required a fair amount of effort: the relation between the solutions for $\nu + \mu_3$ belonging to a solution E and its Ehlers transform $\tilde{E}$ had to be established: the equations governing q_2 had to be integrated (which required some ingenuity); and the expressions for the Weyl and the Maxwell scalars had to be expressed in terms of the solution of the Ernst equation for H. The last reduction was essential since it turned out that the solution obtained by applying the Ehlers transformation to $E = p\eta + iq\mu$ was of type-D — a fact which could easily have escaped notice.

The solutions that were obtained from $\tilde{E}$ were similar to the one obtained in the paper on the 'New Type of Singularity' in that an event horizon was formed with a subsequent development of a time-like singularity along hyperbolic arcs.

After completing the discussion of the Ehlers-transformed solution, we returned to the general solution for the hyper-surface orthogonal case (i.e. when $q_2 = 0$). The solution obtained in this case provided a generalization of the Bell–Szekeres solution in the same way that the solution for static-distorted black holes provides a generalization of the Schwarzschild solution. The space-time develops a horizon and subsequently a *three-dimensional* time-like singularity. In the course of this work, Basilis and I developed some difference of opinion with regard to the usefulness of obtaining the explicit solution for $\nu + \mu_3$. The matter was put aside even though I had, in principle, succeeded in solving the problem. The details still require to be worked out.

There was a minor digression that we undertook: to find the Einstein–Maxwell analogue (for the case $Z = 1$) of the Nutku–Halil solution for the vacuum.

All of these were completed by the end of June; and the paper was written and sent to the Royal Society on July 17 prior to my visit to the Bell Laboratories on July 18.

At long last, I could return to thinking about my Schwarzschild lecture. As I have said, I had decided to make a special effort for this lecture. The subject that seemed most appropriate was 'the aesthetic base of general relativity' — a topic that I had wished to explore for a year and more. There was not much time left to think since we had to prepare for the Bayreuth Festival — reading the librettos of the seven operas (the four operas of the Ring, The Meistersinger, Tannhauser, and Tristan and Isolde) and hearing them on records (which we had acquired for the purpose). Still I got myself into the proper frame of mind by writing a critical essay on the views that Dirac had expressed in his UNESCO lecture on the 'Excellence of General Relativity', and compiling some account of Schwarzschild's three papers — on star streaming, on the radius of curvature of the three-dimensional space of astronomy, and on his solution of Einstein's vacuum equations.

Only after returning from Bayreuth, could I turn my full attention to the preparation of the Schwarzschild lecture. I had barely three weeks; and I had to allow for Lalitha's second eye-surgery during the same three weeks.

The writing out of the Schwarzschild lecture was a major effort: I had to explore in depth the 'aesthetic base' of the general theory of relativity which I had never done before. (In this connection, my earlier discussions with Wald and Basilis on my essay on Dirac's views were helpful). But the most difficult part was to explain how sensitiveness to the aesthetic qualities of a theory can enable one to formulate and solve concrete problems that lead to a deeper understanding of the physical content of the theory. The only example that I could describe in detail in this connection was the construction of the theory of colliding waves by patterning it after the theory of black holes. A chart illustrating the similarity in the two patterns was in many ways the heart of the lecture, though I am afraid that none of my listeners at the lecture (nor the readers of the printed version) will follow these parts of the lecture.

The unexpected discovery of Einstein's *Gedllchtnisrede* for Schwarzschild at the Berlin Academy in June 1916, required a 'last minute' effort to translate it for inclusion as an Appendix to the Lecture. (I had the translation checked by D. Mülller.) The result was that the complete text of the lecture could be ready only by noon just about five hours before the scheduled departure of my plane to Hamburg on September 17.

October 9, 1986

At Hamburg, I had the long awaited opportunity to see and to talk to Lo Woltjer at leisure in a relaxed atmosphere. We had dinner together that evening; and on the following morning, before my lecture, we went for a long walk in the garden around the University talking about many things: his administrative experience and how we looked at our futures — very different but with understanding.

As for the lecture itself, I am afraid the audience — typically astronomical — was uneducated in relativity. Even Weigert (who played as my host) said he should have to read my lecture in print before he could express an opinion. I had expected that the response would be slight: but I did not want to 'dilute' the lecture and was steadfast in my determination to do justice to Karl Schwarzschild's legacy.

Next morning I flew to Leiden, having accepted an invitation from van der Laan. It gave me the opportunity to meet Jan Oort and Henk van de Hulst. The difference in outlooks engendered over the years had widened too greatly for a genuine meeting of the minds.

And my talk on colliding waves simply passed them by (how different was the reception to essentially the same talk at the Raman Institute in Bangalore two months later!).

Returning to Chicago on September 20, I faced an extremely heavy schedule before leaving for India on December 9.

Valeria arrived on September 22 to complete the work on the dispersion of cylindrical impulsive waves that we had started during her earlier visit (May 5–31). I was doubtful of the outcome and was afraid that the investigation may be protracted and inconclusive. Besides, there were lectures to give: at Bard College (October 24), Notre Dame (October 31), and T. D. Lee's Symposium in Columbia (November 22).

Soon after Valeria came, I realized (and discussions with Narasimhan confirmed) that the discontinuous integral we had chosen was divergent and had to be abandoned. The convergent integral that we next considered did represent an impulsive wave — but impulsive in the manner of sound waves emitted by a struck string, the time-derivative having a δ-function behavior. Valeria showed that equations governing the C-energy could be integrated. She preferred to work with the solution expressed in terms of the hypergeometric function. I preferred the solution expressed in terms of the elliptic integrals: it had the advantage that explicit expressions for the relevant quantities including the Weyl scalars could be found. The determination of the behaviors of the various quantities at the discontinuities was a delicate matter. But Valeria's experience with computing enabled the checking of the various formulae against numerical evaluations. Valeria left on October 17; but it took all of November to resolve the remaining discrepancies. With persistence the paper was written, all the illustrations were done, and the paper sent to the Royal Society on December 4.

With the paper sent, I still had to write the foreword to my collected essays; and the tribute to Ambarzumian on his 80th birthday, leaving only the paper that I was to write with Basilis untouched.

We left for India on December 9, arriving in New Delhi very early on the morning (1 a.m.) of December 11. The Vainu Bappu Lecture at INSA was to be given on the following day.

On the morning of December 12, we visited the University of Delhi; and on the way stopped at Kothari's house. We had an exceptionally cordial meeting. Giving the lecture at INSA was not a satisfying experience: with the large audience assembled, I had to be so 'popular' that nothing significant could be said. Even so, Kothari was sufficiently interested to ask me to send him my reprints on colliding waves.

On the next day we visited the U.S. Embassy for a luncheon meeting with Ambassador Dean and some Indian scientists (including G. K. Menon, C. N. R. Rao, D. S. Kothari and A. P. Mitra).

After Delhi we had a brief vacation at Jaipur and Udaipur — the visit to the Jain temple with 400 marble pillars was a memorable one. Then Bombay, lecture at the Tata Institute, meeting with the Relativity group and the mathematicians. Then a week in Madras and ten days in Bangalore. The stay in Bangalore was a most rewarding one: the relaxed atmosphere at Balakrishnan's home, the attentiveness and hospitality of Shyamala, and the visits of Savitri and Vidya before we left. The only regret was that the visit was short; and the days went by leaving a gap at the end.

At Bangalore, I gave a seminar-lecture at the Raman institute to a small group surveying the work on colliding waves.

So back to Chicago and to work. The first and overriding task was to find out what Basilis had done with respect to the Einstein–Maxwell analogue of the solution obtained in our paper "A new type of singularity created by colliding gravitational waves." The expectation was a solution belonging to type-D and an extension leading to a time-like singularity along a hyperbolic arc. Basilis had left one question unresolved: the proper choice of a constant of integration for q_2. The choice depends on the context: its vanishing for $\eta = 1$ for a convenient continuation beyond $s = 0$ and its continuity when $q_2 = 0$ for local isometry with the Kerr–Newman solution. Basilis also had investigated (as per our earlier agreement) the behaviour for $s = 0$ of the solution we had obtained in our earlier paper on the Einstein–Maxwell analogues of the Nutku–Halil solution. I did not like the indirect method followed by Basilis. I investigated the matter *ab initio* by a direct method which is the one described in the published paper. And finally, I found that the hyper surface orthogonal case had to be investigated separately; in contrast to the solution for the vacuum, the solution develops a horizon reminiscent of the Reissner–Nordström space-time.

By the time I had completed my analysis of the Einstein–Maxwell case, Basilis had arrived. He had expected to complete the entire paper in three weeks. But that was impossible because of the additional material to be checked and verified already in the Einstein–Maxwell case. Taking up the $(\varepsilon = p)$-fluid with the type-D vacuum solution, I found many gaps to fill. The required spin-coefficients for the vacuum had to be evaluated; and all the asymptotic behaviors established. When Basilis left on February 18, the first two parts of the paper had been written in final form.

After Basilis left I took up the third part that was to deal with the null dust. Apart from completing the 'skeleton' that Basilis had left, I decided to go in depth in the comparison of the solutions for the $(\varepsilon = p)$-fluid and the null dust with the corresponding solutions in special relativity. This proved not to be as straightforward as I had thought (there is in fact an error in Penrose's analysis not 'disclosed' in the published paper). And I also had to resolve some of Basilis's doubts and misunderstandings. With all these additional complications, the paper was finally completed only on March 11 and sent in to the Royal Society. (But some changes were made a week later.)

With the completion of the paper, I could finally turn to the Principia. I first refreshed my memory of the circumstances that led to the writing of the Principia. I found Rouse Ball's Essay most useful. I did not find that Westfall served my purposes. After some careful thought, I decided to select some ten propositions out of the Principia, write out my own proofs, and compare them with Newton's. The task was more time consuming than I had thought. All of March and April were taken with my preparation for the lectures I gave on the campus and in Washington. I have a fairly complete manuscript. But this will have to be rewritten in the summer for publication.

May 3, 1987

S. Chandrasekhar

PAPERS:	Communication date	Acceptance date
1. A new type of singularity created by colliding gravitational waves (with B. Xanthopoulos), *Proc. Roy. Soc. Lond. A* **408** (1986) 175–208.	7 April 1986	14 April 1986
2. Cylindrical waves in general relativity, *Proc. Roy. Soc. Lond. A* **408** (1986) 209–232.	24 April 1986	1 May 1986
3. On colliding waves that develop timelike singularities: a new class of solutions of the Einstein–Maxwell equations (with B. Xanthopoulos), *Proc. Roy. Soc. Lond. A*	17 July 1986	21 July 1986
4. On the dispersion of cylindrical impulsive gravitational waves (with V. Ferrari), *Proc. Roy. Soc. Lond. A*	4 December 1986	9 December 1986
5. The effect of sources on horizons that may develop when plane gravitational waves collide (with B. Xanthopoulos), *Proc. Roy. Soc. Lond. A*	11 March 1987	17 March 1987

LECTURES

1. The aesthetic base of the general theory of relativity (Karl Schwarzschild Lecture), Mitteilungen der Astronomischen Gesellschaft Nr. 67 (1986).

TRIPS

1986:

27 January, Oxford, England, discussions with Roger Penrose

13–14 March, Munich, Germany; March 15-22, ESO/CERN, Geneva: Colloquia

21 April, Wayne State University, Detroit, Michigan. Vaden Miles Memorial Lecture

1 May, Northeastern University, Boston, Massachusetts. Eighth Distinguished Scientist Lecture

30–31 May, Princeton University, Princeton, New Jersey. Martin Schwarzschild Birthday Celebration

18 July, AT&T Bell Laboratories, Murray Hill, New Jersey. Colloquium

18–30 August, Bayreuth, Germany, Wagnerian Festival

17–20 September, Astronomische Gesellschaft der Virsitzende, Tubingen, Germany. Karl Schwarzschild Lecture. Visit to Leiden Observatory, Leiden, Holland. Colloquium.

25 October, Bard College, Annandale-on-Hudson, New York. Distinguished Scientist Lecturer

31 October, Notre Dame University, South Bend, Indiana. Distinguished Physicist Lecturer

21–22 November, Columbia University, New York, New York. Symposium in celebration of the 60th birthday of T. D. Lee

12 December–9 January (1987), Indian National Science Academy, New Delhi, India. Vainu Bappu Award

1987:

21–22 April, Washington, D.C. Symposium on the History of Astrophysics, American Physical Society

22–25 April, University of Maryland, College Park, Maryland, Celebration of the Tercentenary of the Publication of Newton's Principia

1986:

1 January–31 August, Xanthopoulos comes to Chicago as Visiting Scientist

5–31 May, Valeria Ferrari

22 September–17 October, Valeria Ferrari

1987:

2–18 February, Xanthopoulos

2–28 February, Ferrari

Continued Efforts II
(May 1987–September 1989)

To take up my "Continued Efforts" where I left off in May 1987, I am afraid that it will be difficult for me to conform to the style of my earlier installments. As will be evident from the appended brief chronology, my time had been interrupted by so many conflicting commitments that it was impossible to follow my normal procedure of concentrating on a particular piece of investigation till its completion before taking up another, I shall therefore give an account of the manner in which the different papers published during this period came to be,

1. To start with the paper "On Weyl's solution for space-times with two commuting Killing-fields" (*Proc. Roy. Soc. Lond. A* **415** (1988) 329), the initial idea for it came during the early spring of 1986 while I was working, together with Xanthopoulos, on our paper "On colliding waves that develop time-like singularities: a new class of solutions of the Einstein–Maxwell equations" (*Proc. Roy. Soc. Lond. A* **419** (1987) 311). In that investigation, I confronted for the first time the problem of solving Equations 133 and 134, which are at the base of completing Weyl's solution for hypersurface orthogonal solutions for space-times with two Killing-fields. Already at that time I

had found the explicit solution of Equation (40) for the "linear part" of the equations. I also realized that, to solve Equation (41) for the quadratic parts, one must make use of the functions introduced by Franz Neumann in 1878. Basilis was not interested in completing the solution. He felt (as he said) "demoralized by having to direct his energies in a futile direction." I therefore put the problem aside in order to complete the investigation on hand. And while I continued to think about the problem at odd times, there were other matters pressing on me: the preparation for the Schwarzschild Lecture in the fall of 1986. And when that was over, I had the task of completing a protracted investigation on the dispersion of cylindrical waves, together with Valeria Ferrari. ("On the dispersion of cylindrical impulsive gravitational waves," *Proc. Roy. Soc. Lond. A* **413** (1987) 75). And by the time that was completed, we had to leave for India.

On our return from India, my principal problem was to complete the paper "The effect of sources that may develop when plane gravitational waves collide" (with Xanthopoulos, *Proc. Roy. Soc. Lond. A* **414** (1987) 1). I have written about this already in my previous installment. And when that was out of the way, I had to work on Newton's Principia for my lectures in Washington, D.C. and in Chicago. And so it was only in May of 1987 that I could return to the problem which I had abandoned for about a year. My initial reaction was that the solution to the problem could be completed in a relatively straightforward way (in the manner of the solution of the linear part) by obtaining simple recurrence relations by making use of Neumann's functions Y_{nm} and Z_{nm}. While I could get, without too much difficulty, one set of recurrence relations, I was misled for a time into believing that that would complete the solution to my problem. I soon realized that it did not; and that another set of recurrence relations was needed. I discussed the problem with Lebovitz and Narasimhan but to no avail. I had to leave for the conference in Cambridge with the problem unsolved.

Returning from Cambridge, I could at last spend a few weeks with undistracted attention to the problem. And that was all that the problem needed! I soon found a way of establishing the second set of recurrence relations and, with some discipline; I was able to derive all the relations listed in pages 345–347 in the published version of the paper. At long last, more than a year after I had started thinking of this problem, the paper was completed. The referee of the paper thought that the most interesting part of the paper was the concluding sentence, "Franz Neumann may have been pleased by the resurrection of his functions, Y_{nm} and Z_{nm} in the context of Weyl's solution!"

2. I shall take next the second paper on the list relating to the perturbations of the Bell–Szekeres space-time. During Basilis visit in February 1987, we had vaguely thought about analyzing the perturbations of the Bell–Szekeres space-time. Some days later Basilis brought his reduction of the Ricci identities. By examining them (Equation (21) of *PRS A* **420** (1988) 93), I was able to decouple them by introducing potentials in the manner described; and these reductions led to Equations (30) and (31). But I did not think further about the separation of these equations at that time; and when Basilis left (February 19) the equations had not been separated.

After completing my paper on the Weyl solution in August, I turned to the separation of Equations (31) and (32). And it was not long before I was able to accomplish the separation and reduce the equations to spin-one weighted spherical-harmonics (Sec. 4a). But one of the separated equations (that for g) involved spin-weighted harmonics for a complex "m". I was fairly convinced that this fact would not lead to any real singularities. It seemed inconceivable that the solutions would have a behavior different from what Hobson had established for the Legendre functions. I thought that this matter could be settled easily and the paper completed without too much additional effort. With this in mind, we went to Crete for a week in late September–early October. Though this week was ostensibly for a vacation, we did work hard, since Basilis felt that the perturbation analysis for Region II should also be carried out. We wrote down the basic equations and by the end of the week we knew pretty well how these equations could be solved. But I was not able to convince Basilis that the spin-weighted harmonics for the complex argument did not lead to any essential singularity. With the investigation in this unsatisfactory state we returned to Chicago. But soon after our return, a long telephone conversation with Saul Teukolsky made it clear that the spin-weighted harmonics could be reduced to the Jacobi polynomials and establish the behavior of these harmonics; and the reduction showed that the behavior was the same as for the Legendre functions. I had left the reduction of the perturbation equations to Basilis. From what I was able to gather on the telephone, his results were not conclusive; and I suggested that he visit Chicago for at least some six weeks during the winter of 1988.

A week after Basilis arrived on March 1, it became clear to me that the space-time in Region II did not allow any non-trivial u-independent perturbations to which we had restricted ourselves. It took about three weeks before it could be established beyond all doubt that Region II did not in fact allow any non-trivial u-independent perturbations. Since neither of us could come to a reasonable understanding of this result, I decided to go to Oxford to consult Roger Penrose. This I did during March 24–27, 1988. Roger seemed to think that it was improper of us not to have included u-dependent perturbations. Arriving at Heathrow next morning, I found that the plane had been delayed by some six hours. I used this time to develop the equations governing u-dependent perturbations. The basic equations were solved on the flight to Chicago.

In the following days the solution was completed; and we found that the u-dependent perturbations diverged in Region II and did not provide the basis for a satisfactory extension. The matter had to end there. With effort the paper was written and completed by the time Basilis left on April 13. Now I turned to the two-centre problem.

3. In spite of the many distractions during the past two years, the problem which was constantly on my mind was the two-centre problem in general relativity. While I had been interested in the Majumdar–Papapetrou solution of the static arrangement of an arbitrary number of extreme Reissner–Nordström black holes, ever since I simplified the extant analysis in writing the last chapter of my *Mathematical Theory of Black Holes*, my active interest was stimulated by a conversation with Gary Gibbons when I happened to sit next to him at the banquet during the Cambridge Conference in July 1987. Gibbons asked me if I had been interested in the Majumdar–Papapetrou solution. When I said I had not, he stated that the solution representing the assemblage of extreme Reissner–Nordström black holes had similarities with solutions representing assemblages of magnetic monopoles. This conversation, together with the somewhat oblique reference to the two-centre problem in the Principia, stirred my interest sufficiently that I wanted to go deeper into the problem of two extreme Reissner–Nordström black holes placed on an axis of symmetry. But before I started a more detailed probing of this problem, I tried näive approaches to the stability of the general Majumdar–Papapetrou solution along the lines I had originally followed while writing my account in *The Mathematical Theory of Black Holes*, i.e. considering $\Psi + \nu$ to be small instead of zero.

I pursued various trails along these lines during the summer and fall of 1987 but without success. I profited by many conversations with Curt Cutler, who, with admirable independence, disagreed, with reason, with many of my approaches. As a result, I decided on a frontal attack on the problem of the axial axisymmetric perturbations of two extreme Reissner–Nordström black holes on the axis. During the fall and in the months before Christmas, I developed the perturbation analysis in spite of the constant distractions with the ongoing investigations of the Bell–Szekeres space-time. Of course I could not continue very long because of the interruption of the visit to Crete, and the preparations for the talk I had to give at the Ramanujan Centennial Symposium in Madras, and of the sub-

sequent visit to Paris. In any event, by the time Basilis and Valeria
arrived in Chicago in March, I had progressed sufficiently far to have
established the conservation theorem. In between all the other distractions, I continued to think about the asymptotic behaviors of the
solutions at infinity and at the two singularities. The problem with
respect to the behavior at infinity was resolved without too much
difficulty, even though the matter was not clarified completely until
much later in the summer. The behavior near the singularities gave
very much more difficulty. I did succeed in decoupling the equations
at the singularities. But the complete resolution of the problem had
to wait. Besides, since the entire objective of the investigation was
to describe the scattering process via a scattering matrix, I realized
that I must learn more about multi-channel scattering. By some
good fortune, I dropped in to Roland Winston's office to find out if
he could enlighten me on these issues. I was delighted not only to
learn that he was in some ways an expert on these matters, but that
unlike most experts, he was willing to take the time to educate me!
There were many pitfalls that had to be overcome. The major obstacle was to realize that the gravitational and the electromagnetic
waves were coupled in such a way that, near the horizons, one had to
think of the radiation field as photon-graviton waves — a surprising
conclusion. Winston and I were so intrigued with this association
that I explained and discussed this matter with Cronin and Rosner.
But substantive remarks relative to this association were made only
by Nambu (who continued to be a consultant on these questions).
However, many strands of the problem remained unresolved even by
the time we went to Lindau. A wrong sign in writing the divergence
relation was an irksome error. Bob Wald located it to my considerable chagrin! After a concentrated effort of some two weeks upon our
return from Lindau, I was satisfied with the outcome of the investigations and the paper was written and sent to the Royal Society in
July.

Already during the spring I had suggested to Valeria to
look into the polar perturbations of the two-centre problem. The

problem was not as straightforward as I had thought; and the subject was left unattended since the simpler problem of axial perturbations remained unresolved. And when that was completed I became interested in another aspect of the two-centre problem.

4. A problem that had intrigued me during the months I was working on the two-centre problem was the following: With two extreme Reissner–Nordström black holes on the axis, we have an axisymmetric solution of the Einstein–Maxwell equations; and underlying the solution was a three-dimensional Laplace's equation. The question that intrigued me was, where does this Laplace's equation come from; and what is its relation to the two other Laplace's equations and the other equations (e.g. the Ernst and the X- and Y-equations) which characterize static and stationary space-times? Besides, I thought that one could obtain the solution for two colliding Reissner–Nordström black holes by some simple generalization (by letting, for example, $f \neq 0$ and $f \equiv f(t)$). I tried several generalizations and at one point I indeed thought that I had found such a generalization. But they all collapsed: they were found to be inconsistent with one or another of the field equations. The moral once again was, one cannot make discoveries by hoping that random thoughts will succeed. I returned once again to the basic question in a systematic way. Writing the metric in prolate spheroidal coordinates, I tried to find other solutions via the conventional methods based on reducing the equations to the Ernst equations. But all these efforts led me away from the understanding that I was seeking. Gradually it became clear that I ought to reduce the equations to a form in which the solutions for the electrostatic potential and for f became manifestly trivial solutions of the underlying equations. Trying to find out how this can be achieved, I came to the rather surprising conclusion that the underlying Laplace's equation was none other than what the X and Y equations become when one or the other is 0! Since I knew that the X and Y equations are basic to the solution of stationary axisymmetric vacuum space-times, the one-to-one correspondence between stationary vacuum solutions

and static Einstein–Maxwell solutions became clear. When I talked to Basilis over the telephone about this correspondence, his initial reaction was that I was discovering what he chose to call 'Bonnor's transformation'. It took some effort for him to realize that the one-to-one correspondence is much deeper than anything that had been thought of before in this connection.

Once the correspondence had been established, the question that immediately sprang to my mind was the following: What is the solution of the Einstein–Maxwell equations that follows from the simplest solution of the X and Y equations that I had derived some ten years earlier? As soon as the metric was written down, it became clear that it represented two charged black holes. I first thought that the entire space-time was smooth. When I told this to Bob Wald, he directed my attention to a paper by Peter Ruback, who had shown that such multi-black hole solutions were impossible. It was clear that the space-time that I had found must violate the smoothness requirement in some way; but the question was, in what way? I had sent my metric describing two black holes to Basilis; and he thought that the solution was characterized by curvature singularities. On that account I lost interest in the solution; and when Basilis came to Chicago in November, he told me that he had found a mistake in his calculations and that what remained was only a conical singularity; and so with renewed interest we examined all the properties of the solution and found, to our surprise, that the upper limit M for the charge $|Q|$ was also violated. It was clear that we had a solution of some physical interest. Discussions with Nambu fully confirmed this impression: charged black holes have properties in the classical domain very similar to properties of magnetic mono-poles in the quantal domain. Since we felt that the solution was of more than normal interest, we worked hard to complete the papers before the Christmas holidays. And we did: the completed papers were in fact sent to the Royal Society the week before Christmas.

January 17, 1989

5. During Valeria's visit during March and April of 1988, I had asked her to start work on the polar perturbations of the two-centre problem while I was still unraveling the conceptual issues in the context of the simpler axial perturbations. I had thought that the part up to and including the conservation theorem would be straightforward. It turned out to be otherwise. Soon enough she derived the required perturbation equations. But all efforts to obtain the flux-integral failed: the equations presented no clear symmetry. Also, I was not able to spend much time with her: I was too occupied with the Bell–Szekeres perturbations; and new conceptual issues continued to arise in the context of the axial perturbations which were eventually resolved only in June and July. But it did occur to me that one should be able to obtain the flux-integral for the polar perturbations by a general algorism. It seemed to me that the Landau–Lifshitz pseudo-tensor might provide a means. So when Valeria left, I suggested that she obtain the components of the pseudo-tensor for the fully time-dependent static axisymmetric space-time. She did send me her results $-gt^{ok}$ $(k = 0, 2, 3)$; and since I was unable to suggest what she should do beyond that, I asked her to visit Chicago for some five to six weeks when that was possible. And it was decided that she would come early in the new year. I felt that her visit in January–February should be more successful than her earlier visit. So when the two papers had been sent to the Royal Society the week before Christmas, I began to think seriously about the problem of how one is to obtain the flux-integral from a consideration by the Landau–Lifshitz pseudo-tensor.

I began by reducing the expressions of $-gt^{ij}$ — all of them. This I did before Valeria came in January. At that time I had no clear idea as to how one should proceed. Meantime, I asked her to verify the linearized version of the divergence relations in the context of the Schwarzschild and the Reissner–Nordström perturbations. The verification with respect to the Schwarzschild background was accomplished with some difficulty. But the verification with respect to the Reissner–Nordström perturbations led to serious discordances:

the terms derived from $-gT^{ij}$ were coming out all wrong. First, by using the expression for $(-gT^{ok})_{,k}$ as a linear combination of T^{ij}'s we showed that there had been some errors in the evaluation of these terms and, more importantly, that the transformation to the tetrad components should not be made after the differentiations. Eventually the linearized version of the equation $\theta^{ok}_{,k} = 0$ was checked. The question of how the flux-integral was to be obtained remained.

First, it was not clear to me how Habisohn had derived the correct conserved energy-momentum theorem for perturbed space-times. Wald explained the procedure to me. The procedure was to substitute formally for the metric coefficients ν, Ψ, etc., $\ldots \nu + \lambda\delta\Psi$, $\Psi + \lambda\delta\Psi$, etc. where ν, Ψ, etc., were time independent and expand θ^{ok} to the second order in X and equate the terms of $O(\lambda^2)$. But I was blocked as to what to do afterwards. I knew that $\delta\nu$, $\delta\Psi$, etc. should be replaced by the linear perturbations with the time dependence $e^{i\sigma t}$; but what then? Rafael Sorkin was fortunately visiting Chicago at that time. He called on us one evening and I talked to him about the impasse that I had reached. After he left, I realized that his remarks (deep as always!) amounted to the suggestion that I equate the time-independent terms in the bilinear expressions that one obtains for the terms of $O(\lambda^1)$. This suggestion was the essential key to the problem. The verification with respect to the Schwarzschild perturbations was easy enough. But the matter was not so easy with the Reissner–Nordström perturbations. After several misunderstandings and detours, the verification was accomplished — largely due to Valeria's persistence.

So after six weeks of strenuous efforts on both our parts, we finally succeeded in obtaining the flux-integral for polar perturbations. Valeria left before the $(n-1)$-version of the paper was written. But it was all completed a fortnight after she left. The paper was written and sent in to the Royal Society on March 3, 1989.

March 22, 1989

The euphoria in having completed the paper was not to last long: an incredible oversight vitiated the entire analysis. But the oversight was noted only a month later; and the paper was eventually withdrawn. I should first explain why.

I was due to give the first Seferis Lecture in Athens on March 29. The lecture had to be prepared carefully: the occasion was an 'important' one, and there was hardly enough time.

They wanted a lecture on a general topic related to my Ryerson and other lectures. Instead of repeating one of the earlier lectures, I decided to change the emphasis slightly to "The perception of beauty in the pursuit of science". And this same topic was to suffice for my lecture to the American Academy of Arts and Sciences to be given later in May.

We left for Athens on March 27. A late arrangement was that after the lecture in Athens, we would spend two weeks as a semi-vacation in Crete and then a brief visit to Spain — to Barcelona and Granada — the latter to be present at the symposium in honor of Guido Münch.

We were shown exceptional hospitality in Greece and in Spain. We enjoyed our vacation in Crete and our visit particularly to Granada. We returned to Chicago on April 22.

Soon after our return from Spain, I realized that in verifying the flux-integral for the Reissner–Nordström space-time, we had overlooked that the final result was wrong by a factor Δ/r^4 our calculations had shown that:

$$\int_{-1}^{+1} E_r d\mu \propto r^2 [Z, Z^*]_r .$$

whereas it should have been

$$\int_{-1}^{+1} E_r d\mu \propto [Z, Z^*]_{r_*} = \frac{\Delta}{r^2}[Z, Z^*]_r .$$

I had become suspicious since Wald had pointed out earlier the possibility of such a factor having been overlooked. My first reaction was that we had made a simple oversight. But that was not the case. Then my feeling was that there *must* be a simple explanation for the additional factor. Could the factor arise, for example, by using the Einstein-complex instead of the Landau–Lifshitz? But that was not the case either. Discussions with Sorkin, Wald and Geroch proved to no avail. Meantime, it had appeared that the flux-integral for the Einstein–Maxwell space-time might be in error more seriously than by a simple factor. Faced with this predicament, I asserted to Wald, with some bravado, that I could after all derive what I wanted *ab initio* directly from the linearized equations. Under the circumstances, there was indeed no other choice! The prospect was appalling; and I had to brace myself to embark on what appeared to be a long and thorny trail — a 'supreme last effort' — as I told myself.

But some encouragement came my way, when I realized (to the surprise of both Wald and Burnett) that the initial-value equations simplified the flux-integral for the vacuum considerably. That turned out to be a key factor.

Again I had to postpone getting into grips with the problem since I had to think about the lectures I was to give at the American Academy and at the Gibbs Symposium. I started to grapple with the linearized field-equations in earnest only towards the end of May.

First, there was the problem of writing the linearized equations in a form that will manifest their internal relationships. The choice of the equations for $\delta\mu_2$ and $\delta\mu_3$ was the key to the entire analysis. The transformations used in passing from Equations (19) iii + iv to (28) and (29) (in the published paper) were essential steps that were arrived at only slowly. An equally important identity that proved crucial is that given in Equations (40) and (41). In this manner the essential ingredients for the derivation of the flux-integral for the vacuum were isolated. The analysis for the Einstein–Maxwell space-time readily followed.

I completed all this analysis by the time Valeria came on July 8 after the General Relativity conference at Boulder.

While Valeria was checking the analysis, I started on the analogous reductions appropriate for the non-radial oscillations of a spherical star. Again a simple oversight — the same one I had made some years ago in connection with the paper on null dust — delayed the completion. But all the loose ends were tied and the final paper was written and sent on August 17. Meantime, we had started on a reformulation of the problem of the non-radial oscillations of a spherical star. But of these, at a later time.

September 5, 1989

TRIPS

1987:

20–22 April: Washington, D.C.: Symposium on the History of Astrophysics, The American Physical Society, Crystal City

22–25 April: University of Maryland Center for Renaissance and Baroque Studies, Symposium on Newton's Principia. "Scientific Creativity"

10 May: Syracuse, N.Y.: Award of honorary D.Sc., Syracuse University

15 May: Milwaukee, Wisc.: Discussions with John Friedman

1–5 June: Urbana, 111: University of Illinois at Champaign-Urbana. Ramanujan Centennial Symposium

29 June–9 July: Cambridge, England: Tricentennial celebration of the publication of Newton's Principia, Cambridge (Trinity College). 'The aesthetic base of general relativity'

17–18 September: Socorro, N.M.: Visit to the Very Large Array (VLA) at the National Radio Astronomy Observatory

27 September–3 October: Iraklion, Crete: Discussions with B. C. Xanthopoulos

21–22 October: Edmonton, Alberta: Theoretical Physics Institute, University of Alberta. Distinguished guest speaker at Symposium on Tercentenary of Newton's Principia, "On the collision of gravitational waves in general relativity"

4–5 December: New York, N.Y.: Columbia University: Board of Trustees of the Taraknath Das Foundation presents the Taraknath Das Foundation Award

22 December–3 January: 23–26 December: Madras, India: Ramanujan Centennial Celebrations; 26 December–3 January: Visit to Bangalore

Papers submitted to *Proc. Roy. Soc. Lond.*:

11 March: The effect of sources on horizons that may develop when plane gravitational waves collide (with B. C. Xanthopoulos)

3 August: On Weyl's solution for space-times with two commuting Killing fields

Papers published:

On colliding waves that develop time-like singularities: a new class of solutions
of the Einstein–Maxwell equations (with B. C. Xanthopoulos), *Proc. Roy. Soc.*
410, 311–336

On the dispersion of cylindrical impulsive gravitational waves (with V. Ferrari),
Proc. Roy. Soc. **412**, 75–91

The effect of sources on horizons that may develop when plane gravitational
waves collide (with B. C. Xanthopoulos), *Proc. Roy. Soc.* **414**, 1–30

Book published:

Truth and beauty: aesthetics and motivations in science (University of Chicago
Press)

Visiting Scholars:

2–28 February: Valeria Ferrari

2–19 February: Basilis Xanthopoulos

Awards:

10 May: D.Sc, honoris causa, Syracuse University

4 December: Taraknath Das Award, Columbia University

University of Chicago Lectures:

14 April: University of Chicago Newton Tricenteiinial Committee (J.W. Stigler,
Chairman) invited lecturer

6 May: Lecture on Newton for the University of Chicago Library Society

TRIPS

1988:

18–22 January: Meudon (Paris), France: Meeting of Nobel Laureates, sponsored
by Prime Minister Mitterand & Elie Wiesel

24 February: Madison, Wise: Discussions with Richard Askey

24–27 March: Oxford, England: Discussions with Roger Penrose

13–14 May: Washington, D.C.: Editorial Board Meeting of the American Scholar

22 May–3 June: 22–27 May, Kazimerz, Poland: University of Warsaw, "New Theories in Physics"

29 May–3 June, Zielonovo: Visit with Andrzej and Roza Trautman

14 June: Montreal, Quebec: Award of honorary LL.D., Concordia University

27 June–1 July: Lindau, Germany: Thirty-eighth meeting of Nobel Laureates. Lecture on "The founding of general relativity and its excellence"

21 August–4 September: Northern Italy and Salzburg, Austria: Vacation in the Dolomites (Italy) and Mozart Festival (Salzburg)

21 October–6 November: Princeton, N.J.: Institute for Advanced Study. The Oppenheimer Lecture: "Newton and Einstein: a study in contrasts"

2–5 December: Owerri, Nigeria: Award of honorary D. Litt. by the Federal University of Technology at Owerri

Papers submitted to *Proc. Roy. Soc. Lond.*:

12 April: A perturbation analysis of the Bell-Szekeres space-time (with B. C. Xanthopoulos)

12 July: The two-centre problem in general relativity: the scattering of radiation by two extreme Reissner–Nordström black holes.

20 December: A one-to-one correspondence between the static Einstein-Maxwell and stationary Einstein-vacuum spacetimes and two black holes attached to strings (with B. C. Xanthopoulos)

Papers published:

On Weyl's solution for space-times with two commuting Killing fields, *Proc. Roy. Soc.* **415**, 329–345.

A perturbation analysis of the Bell-Szekeres space-time (with B. C. Xanthopoulos), *Proc. Roy. Soc.* **420**, 93–123.

Massless particles from a perfect fluid, Nature, 333, 596 A commentary on Dirac's views on "the excellence of general relativity," in *Festi-Val: Festschrift for Val Telegdi*, K. Winter, ed. (Elsevier Science Publishers, North Holland), pp. 49–56.

Visiting Scholars:

1 February–15 May: Valeria Ferrari

1 March–13 April: Basilis Xanthopoulos

1 November–1 December: Basilis Xanthopoulos

Awards:

Honorary Member, American Meteorological Society

14 June: LL.D., honoris causa, Concordia University, Montreal, Quebec

3 December: D. Litt., honoris causa, Federal University of Technology, Owerri, Nigeria

1989:

9 January–18 February: Valeria Ferrari — visiting scholar in Chicago

*3 March: Landau–Lifshitz pseudo-tensor (with Ferrari) (later withdrawn)

27 March–13 April: Athens: First Seferis Lecture (as Fulbright Lecturer), March 29. Crete: Visit with Xanthopoulos, 20 March–12 April

13–16 April: Barcelona, Spain: "How one may explore the physical content of general relativity"

16–20 April: Granada, Spain: "The intellectual achievement that the Principia is" (Guido Münch Lecture)

20 April: Iberia Airlines on strike. Drove to Madrid

21–22 April: Madrid: Lecture, "On black holes"

10–11 May: Cambridge, Mass: Address to the American Academy of Arts and Sciences: "The perception of beauty and the pursuit of science"

12 May: Washington, D.C.: Meeting of the Editorial Board of The American Scholar

14–17 May: New Haven, Conn.: Josiah Willard Gibbs Lecture, Yale University: "How one may explore the physical content of general relativity"

8 July–17 August: Basilis Xanthopoulos (visiting scholar) in Chicago

8 July–30 August: Valeria Ferrari (visiting scholar) in Chicago

*17 August: The flux integral for axisymmetric perturbations

17 August: Withdrew Landau-Lifshitz pseudo-tensor

Valeria Ferrari working with Chandrasekhar

Continued Efforts III
(September 1989–October 1991)

1. The non-radial oscillations of stars:

My work with Valeria during the past two years has been a chain of trials, euphoria, frustrations, euphoria again, to be followed by humiliation and a final successful denouement which at an earlier time would have provided as much satisfaction as any work that I have done. Now only the humiliation remains.

As I have written in the earlier installment on "Continued Efforts II (May 1987–September 1989)" the completion of our work on the flux integral in August 1989, was preceded by an earlier paper (communicated in March) which had to be withdrawn on account of an obvious oversight which made the entire procedure suspect; and I have described how, on that occasion, prior to Valeria's arrival in Chicago (on July 8, 1989) I had obtained the correct results by an entirely different but direct route. In July while Valeria was checking my calculations, I turned my attention to exploring whether a similar flux integral could be obtained for the non-radial oscillations with attendant emission of gravitational radiation — in effect to an earlier love of mine to which I had paid attention some 18 years earlier when John Friedman and I, in the spring of 1972 while in

Oxford, had studied the equilibrium and the stability of axisymmet-ric systems to axisymmetric perturbations. Valeria and I did obtain a flux integral, that could in principle be used to determine the vari-ation of the flux of gravitational radiation through a star. At that time we had no idea how the flux integral could be applied: we had not even begun our re-examination of the problem of the non-radial oscillations of stars from the point of view of scattering theory.

The derivation of the flux integral appropriate for the polar oscillations of a static star was fairly straightforward. An error in deriving the equation governing the conservation of baryon number (Eq. (108)) was annoying at the time.

During the time I was writing the paper on the flux integral (August 1–20) Valeria continued with writing out the perturbation equations following the treatment of the Schwarzschild black hole.

Towards the later part of August, while examining the equations Valeria had derived, we found to our surprise that the equations al-lowed the integral $\varepsilon N = pL$. I was suspicious of it from the very beginning; and I wished — indeed implored — John Friedman to check our calculations. He never did. Ten months were to elapse be-fore we discovered that the integral simply did not exist. The error arose from an unfortunate confusion in the conventions that were adopted in writing the equilibrium and the perturbation equations: $G_{ab} = +2T_{ab}$ in the equilibrium equations and $G_{ab} = -2T_{ab}$ in the perturbation equations. Fortunately, this confusion in the conven-tions did not affect the flux-integral that we had derived: it was consistent with the perturbation equations as we had written them.

The integral produced a state of euphoria which was to be dashed at a later time. But the equations as derived, though er-roneous, did present a problem that had to be resolved. The same problem is presented by the correct equations. Overcoming it at this time did facilitate the solution of the correct equations at a later time. For this reason, I shall go into the nature of this problem.

What we found with our perturbation equations was that when we attempted to find the behavior of the solutions at the origin $(-r^x)$ via an indicial equation for the exponent x, we obtained the paradoxical result that was undetermined: *every value of x was permissible.* This was an impasse that we did not know how to overcome. We had several discussions with Norman Lebovitz and Persides; but to no avail.

However, after Valeria left, I was able to resolve the problem, with respect to our particular system of equations, in September. The crucial observation was that the system of equations is linearly dependent at the origin. By a careful examination of how this linear dependence came about, I was able to find a linear combination of the equations (allowing differentiation) which was linearly independent at the origin. I could then derive a well-defined indicial equation with the necessary number of distinct roots. (The manner in which a system of linear equations, linearly dependent at the origin, could be manipulated to yield a system which was linearly independent was to provide the key to obtaining the correct set of equations at a later time.)

I did not proceed beyond deriving the indicial equation since the problem that remained was largely numerical. Besides, during September and October, I had a number of other commitments — lectures at Houston, Yorktown and Ottawa and preparing for our trip to India and the lectures on Newton and the Convocation Address at Roorkee.

We returned to Chicago in mid-December (1989), somewhat earlier than we should have liked, since Valeria was to come to Chicago in January and I wanted to prepare for the next phase of our work relating to the solution of the vacuum. However, because of the ill health of her mother and her eventual death, Valeria arrived in Chicago only early in February. Meantime I amused myself by deriving explicit expressions for the Teukolsky–Starobinsky constant for arbitrary spin. (I later communicated the results in the form of a short paper to the Royal Society which was eventually published.)

After Valeria arrived in February, we were principally occupied with obtaining solutions that would describe the interior correctly. And that was not easy since the indicial equation allowed only one singularity-free solution. We were forced to accept a solution with a pole at $r = 0$ satisfying ourselves with the fact that the perturbations for the physical variables were non-singular at the origin — strictly an invalid argument! (I must mention parenthetically, that I was continually disturbed by the fact that the behavior at the origin was not the expected r^ℓ.) The integration of the equations was beset with instabilities; and we sought advice in vain from experts. But in spite of the instability of the integration procedure we thought that we had solved the interior problem adequately. The next problem was to join the interior solutions to the solution for the vacuum and a number of technical problems had to be resolved — but none insurmountable.

Of course, all the time I was constantly worried about the integral on which the work was based. We were totally misled when the numerical integrations provided for the real and the imaginary parts of the frequency of the quadrupole quasi-normal mode values in relatively good agreement with those Lindblom had computed for us by his method. And I am afraid that this "success" resulted in my brushing aside the misgivings that I had felt all along. In any event, in a state of euphoria the paper was written (and I must confess with a tone of arrogance in some parts). The completed paper was sent to the Royal Society a week before Valeria left on March 29.

I had already arranged with Roland Winston during the winter that I would give a series of ten lectures on the Principia during the spring quarter. My first lecture was to be on April 11. I had just about two weeks to start preparing for the lectures. In addition to these lectures on the Principia, I gave the substance of the work on the non-radial oscillations at the Pittsburgh Symposium in honor of Ted Newman. In the subsequent discussion, Kip Thorne raised some questions which were to prove crucial.

The crash came in mid-May when I discovered that the integral on which our work was based simply did not exist: and Bernard Whiting discovered the same thing independently and simultaneously. However, I immediately recognized that the decoupling of the metric perturbations from the hydrodynamic perturbations could still be carried out and that for the barotropic case the resulting equations would be very simple. I informed Valeria of this catastrophe and withdrew the paper I had communicated.

Even though the error that we had made was a humiliating one, I could not spend any time thinking about the resolution of the problem: I was too preoccupied with my lectures on the Principia. But I did tell Valeria how we could resurrect the paper by starting *ab initio* with the barotropic case. Also I should add that I never lost faith in the basic validity of the point of view that we had expressed in the paper.

By the time I had finished my lectures on the Principia, Valeria had tried to obtain the indicial equation for the revised set of equations, and she sent me her reductions. I was not satisfied with the manner of her reductions, but she did find that the indicial equation allowed a double root with the behaviour r^ℓ at the origin. And that was most reassuring.

I was able to take up the problem myself only on June 8, the day after my last Principia lecture. By June 12 I had revised the entire theory and on June 20 I was able to send Valeria the complete basic theory together with the equations necessary for the numerical solution of the problem. Since many of the technical details had been resolved at an earlier time, it was possible to work out the entire theory in essentially a week's time. Valeria at that time was in Spain. She was to return to Rome by the end of June, and was to leave again for her vacation on July 12. I suggested that I go to Rome during July 2–10 and complete the work if possible.

By the time I arrived in Rome on July 2, Valeria had verified my analytical derivations and had started on the numerical integrations. The integrations for the quadrupole mode were completed on Thursday, but on rechecking her calculations, Valeria decided that the integrations should be done differently to achieve the required accuracy — and was confident that she could do it all during the week of July 11 after my departure (postponing her planned vacation).

Then on Friday, we took up the dipole oscillations. When we started the calculations, I did not know what the correct boundary conditions were. I thought that we could discover it empirically, since we had the correct characteristic frequency from Lindblom. First I thought that the relation between L and N (that followed from the vacuum equations) will provide the requisite boundary conditions. But Valeria found by numerical integrations that the relation was identically satisfied — a fact that could be analytically established (by hind-sight!). And only on Sunday evening, after a number of detours following false trails, did Valeria discover the right manner of solving the problem. She worked all of Sunday night, and when she came to pick me up at the hotel on Monday morning, she had the complete numerical solution to the problem, the final result agreeing astonishingly well with Lindblom's. We felt sufficiently elated at this success to spend the afternoon at the Sistine Chapel.

And so when I left Rome on Tuesday morning, all the problems had been resolved: it had taken us only one month.

It took two more weeks to write up the entire paper and send it to the Royal Society on July 20.

2. Non-radial oscillations of slowly rotating stars (Lense–Thirring Effect)

Already when I was in Rome completing the work on the non-radial oscillations, I had mentioned to Valeria that the problem we should next concentrate on was that of a slowly rotating star in which the distortion of the figure $0(\Omega^2)$ can be neglected. After the paper had been sent to the Royal Society on July 20, I started thinking about this problem and came rather hastily to the conclusion that the relevant wave equation could be written down almost at once. But that was muddled thinking, without careful attention to the relative orders of the polar and the axial perturbations. My discussions with Wald and Friedman only tended to confuse the issues further. In any event, I adamantly kept to the belief that the equation I had derived (by illegal elimination from Equations (24), (25) and (28) of the printed paper),

$$(e^{-3\Psi+\nu-\mu_2+\mu_3}X_{,2})_{,2} + (e^{-3\Psi+\nu+\mu_2-\mu_3}X_{,3})_{,3} + \sigma^2 e^{-3\Psi-\nu+\mu_2+\mu_3}X$$
$$= -4[(\varepsilon+p)e^{\nu+\mu_2}\xi_2\varpi]_{,3} + 4[(\varepsilon+p)e^{\nu+\mu_2}\xi_2\varpi]_{,2}$$
$$= [e^{-\Psi-\nu+\mu_2-\mu_3}X\varpi_{,3}]_{,3} + [e^{-\Psi-\nu+\mu_3-\mu_2}X\varpi_{,2}]_{,2}, \tag{1}$$

was correct. I further thought that the numerical integration of this equation could not take very much time. Having sent the relevant formulae to Valeria by the end of August, I suggested that I come to Rome during September 4–12 to complete the work. However, some of the doubts which Valeria had raised about the validity of Equation (1), combined with my own uneasiness, convinced me a day or two before my departure to Rome that the equation was indeed wrong. So when I arrived in Rome on the morning of 5th September I had to tell Valeria on the drive from the airport to the hotel that the development that I had sent her was invalid; and I had to confess that I did not know really how one should proceed. Valeria was, of course, very disappointed.

I rested for an hour or two and I decided that I would not (as I normally do on such occasions) have a shower and go out for dinner; instead, I decided to stay in the room and think about the problem. I thought almost continuously all that night except for brief snatches of rest in between. By breakfast time next morning, I knew precisely how one should proceed.

When Valeria came to pick me up at the hotel at 9 a.m., I could tell her that over the night I had a "revolutionary idea" and that I was certain that it would work.

The "revolutionary" idea was that in the Equation (153) in S. Chandrasekhar and John Friedman; *Selected Papers*, p. 351, we may substitute for $\delta\Psi, \delta\nu, \delta\mu^2, \delta\mu^3, \xi^2$ and ξ^3 (being of $0(1)$) the expressions appropriate for the polar perturbations. Then inclusive of terms $0(\Omega)$ we have the equation:

$$(e^{-3\Psi+\nu-\mu_2+\mu_3}X_{,2})_{,2} + (e^{-3\Psi+\nu+\mu_2-\mu_3}X_{,3})_{,3} + \sigma^2 e^{-3\Psi-\nu+\mu_2+\mu_3}X$$
$$= +[\varpi_{,2}(3\delta\Psi - \delta\nu - \delta\mu_2 + \delta\mu_3)]_{,3}$$
$$-[\varpi_{,3}(3\delta\Psi - \delta\nu + \delta\mu_2 - \delta\mu_3)]_{,2}$$
$$-4[(\varepsilon+p)e^{\nu+\mu_2}\xi_2\varpi]_{,3} + 4[(\varepsilon+p)e^{\nu+\mu_2}\xi_3\varpi]_{,2} \tag{2}$$

This equation couples the axial and the polar perturbations with the coupling provided by w, i.e. by the dragging of the inertial frame.

Valeria and I discussed the problem together all morning and before we went to lunch, we were pretty certain that the basic idea was sound; and the preliminary steps that we should take were clear.

During the following days we worked in parallel, and each of us committed our share of mistakes in our calculations. The principal obstacle in making progress towards the reduction of the equation was the need to find an explicit relation between the Gegenbauer polynomials, $C_{\ell+2}^{-3/2}$ and the Legendre polynomials P_ℓ. The required relation is in fact given in *M.T.* (page 144, Equation (21)). I had forgotten how I had derived it; and I spent one evening rederiving it *ab initio*. Once the exact form of this relation was known, the rest of the analysis was fairly straightforward. And by the end of the week, we had worked out the theory completely and only the numerical work remained. Valeria thought that the numerical work would not take more than a week.

On returning to Chicago, I started to write the paper and found some additional errors in our calculations: but before the end of the week of September 13, the entire theory had been ironed out and we confidently expected that the calculations would be ready by September 20. I should add that during the process of writing, I was able to establish the selection rule $\Delta\ell = \pm 1$ and also the validity of the propensity rule (about which I learned accidentally from Ugo Fano).

My intention was of course that the paper should be sent to press before Lalitha and I were to depart for our vacation in Vienna. But as ill luck would have it, the swelling on my leg, which was increasing (I was, in fact, alarmed by it already in Rome), had become serious enough for me to express my concern to Dr. Sorensen. On seeing him in his office on Tuesday, he shared my concern and suggested my having a complete test for possible clots of blood the following day. There was indeed a spreading of blood clots in many of the veins. But fortunately the spreading had not extended to the main arteries. It was clear that I had to be hospitalized immediately for heparin treatment. At this point it looked as though both our trip to Vienna and the completion of the paper before that, were in jeopardy. The doctor told me that I would be in the hospital for about a week; but I insisted that I must be discharged on Monday and that I would walk out of the hospital without their official discharge if that was necessary! They took me seriously enough to accelerate the treatment; and I was discharged on Monday evening. But during the preceding two days, they had provided me with a desk so that I could draft the paper. It was thus possible to stick to my original plans. The paper was sent in on October 12 and we left for Europe on October 13.

The trip to Vienna and London for the period October 14–28 had been planned, long before, for two reasons: to relax in a different atmosphere after two years of continuous effort and frustrations; but mostly to escape the hypocrisy that will attend my 80th birthday on October 19. And our departure to Vienna on October 13 was none too soon.

We spent ten days in Vienna. We were fortunate with the weather: we could sit out in the sun — in promenades and gardens (e.g. the Belvedire). We attended two operas — Simon Boccanegro and Otello — and two concerts by the Vienna Philharmonic (one of them, played in memory of Leonard Bernstein, included the adagio from Beethoven's 7th Symphony, the Linz Symphony of Mozart, and the Great Symphony of Schubert — altogether a most satisfying program). And in between I reread *The Brothers Karamazov.*

Our four days in England — two in London and two in Oxford — were in the nature of a reunion. On the first evening (Wednesday, October 24) we had dinner with Meggie Weston–Smith in her home. She was writing her father's biography; and we had much information to exchange. On Thursday, besides the Editorial Committee Meeting of the Philosophical Transactions during the morning, David Shoenberg came from Cambridge. We had

dinner together and went to see the production of *Moscow Gold* —
a contemporary political piece with Gorbachev, Yeltsin and others.

The two days with Roger and Vanessa Penrose in Oxford were
very special. Besides being entertained at two dinners — with his
(Roger's) associates in a restaurant and in their new home on the
Oxford Canal, I had a long two-hour discussion with Roger in his
office on Saturday morning. The topic of our discussion was motivations in the pursuit of science and the aesthetic component. Could
one cultivate aesthetic sensibility? Roger described how his motivations operate at two levels: a global and a particular. For example,
his interest in developing the theory of twistors originated in his belief that the theory will provide the basic language for all of physics;
and that is at 'global' level. But it is the development of the theory in
concrete contexts — e.g. the developments due to Woodhouse, Ward
and Atiyah — that provides substance to his efforts. To a question
that Roger asked me concerning my own aesthetic motivations, the
comparison with Monet's serial paintings occurred for the first time.
Having seen only recently an exhibition devoted to Monet's serial
paintings, I could see the relevance of the comparison. The underlying mathematical structures of the theory of black holes, of colliding
waves, and of the non-radial oscillations of the stars are all essentially
the same; but the physical contents are as diverse as one could wish:
like the serial paintings of the grain stacks. The grain stacks are the
same; and the fields on which they are erected and the background
(including Monet's own house) are also the same. But the aesthetic
content of the different paintings is "as diverse as one could wish".
The recollection of our discussions that morning will long remain the
source for further thoughts.

We returned to Chicago on October 28: just a week before
Valeria was to arrive and bring to an end our work on non-radial
oscillations of the stars. The problem that was uppermost in my
mind was the application of the flux integral that we had derived
a year earlier. It was in fact the existence of a flux integral that
suggested, in the first instance, our reformulation of the problem

of the non-radial oscillations of stars. It soon became clear that to apply the flux integral we needed the general complex solution for the quasi normal mode belonging to the complex characteristic frequency $\sigma_0 + i\sigma_1$. And that required a deeper understanding of the Breit–Wigner formula. Discussions with Nambu and Winston were of no avail — at this stage. "If it is not the Breit–Wigner formula, what else can it be?" A "break-through" appeared imminent. But an interruption delayed immediate progress.

I had recommended to John Friedman that he invite Valeria for a Colloquium in order that she may have a chance to talk to him about our work and also obtain some concrete information on neutron stars that may be relevant to our work. She went to Milwaukee on November 16 and gave a colloquium on our latest work on the non-radial oscillations of slowly rotating stars induced by the Lense–Thirring effect. She returned with some 'disturbing questions' that John had asked. Why are the quasi-normal modes dependent on the amplitude of the polar perturbations? Shouldn't they be independent? Besides, would not the polar modes be similarly affected by the axial modes? And would that modify our analysis in any way?

The questions were disturbing enough. But a careful discussion convinced us that while our mathematical analysis was 'flawless,' we had greatly obscured the precise problem to which we had addressed ourselves. A revision was clearly called for. So I called Wainwright at the Royal Society to hold our paper in abeyance as we wished to revise some sections of it; and that he should receive the revisions within ten days. Wainwright told me that the referee of our paper, Bernard Schutz, had volunteered to write to us directly and that we should wait for his letter. Schutz's letter (written on November 19) arrived a few days later.

The revision that was required was to formulate without ambiguity the mathematical problem to which we had addressed ourselves in the context of Equation (2). It then became clear that the parameter in the solution for the quasi-normal modes was none other than w_c (i.e. Ω). Also, the polar modes are indeed affected: the resonant modes already present are affected to $0(\Omega)$. We also revised the two last sections on "an illustrative example" and "concluding remarks." Valeria reconfirmed her numerical results by using some additional test that we had derived (see below): and we also included some remarks on the range of w_c that are relevant. And in the concluding remarks we were more "expansive."

The final revised version was sent to Schutz on December 3 (Monday); and I called (by prior arrangement) on Friday, December 7, to find out if he had further questions. While he agreed that our analytical treatment was indeed flawless, he objected to the use of 'quasi-normal modes' to describe our results. We do not subscribe to his objections: but we added a pro-forma note about his objection. And the revised version was sent on December 7.

During the revision of the paper on the 'Lense–Thirring' effect, I had found that the imaginary part of the complex solution,

$$Z_c = Z + iZ_i\,,$$

was given by

$$Z_i = \sigma_i Z_{,\sigma}$$

while the real part Z is the same solution that we had found for real σ. The result is more than formal. With a careful definition of Z, σ, we were able to show that the minimum of the flux of radiation at infinity, as measured by $(\alpha^2 + \beta^2)$ does indeed locate the real part σ_0 of σ_c. Besides, we obtained an explicit formula for the imaginary part that was new. Valeria verified that σ_i given by the new relations agreed with the value determined by the curvature of the parabola $(\alpha^2 + \beta^2, \sigma)$ at σ_0. A related expression for the Wronskian was also verified. But the numerical calculations had to be carried out with extreme precision which Valeria accomplished.

The relation of this demonstration with the conventional formulation of the Breit–Wigner formula was not clear. Winston clarified the matter and he agreed to write an Appendix to the paper.

Once the complex solution belonging to the complex frequency, $\sigma_0 + i\sigma_i$, was explicitly known, the application of the flux integral to determine the variation in the flux of the gravitational radiation through the star was fairly straightforward. During Valeria's absence in Pittsburgh during the week of December 4, besides preparing the revised version of the paper on the Lense–Thirring effect for re-submission, I drafted the $(n - 1)$ version of the new paper on 'Further Amplifications' we were preparing.

Valeria returned on December 10: and we had exactly one week to complete the numerical work required to illustrate the variation of the flux of radiation through the star. A fact that emerged very clearly from the calculations is that the emission of gravitational radiation by the conventional solution, requires as an initial condition that a δ-function source be provided at $r = 0$.

What remained was to write the final version of the paper, prepare an illustration for the graphic arts, and type the manuscript. All of this, with the exception of the illustration for the printers and Winston's Appendix were completed during that week; and I was able to hand Valeria the final copy just as she was leaving for the Airport (on December 17).

I had arranged to have a bursitis on my foot removed by surgery on Tuesday (December 18).

With the cooperation of Winston and of the 'Graphic Arts' Department, the paper was ready to be sent to the Royal Society on Thursday, December 20.

By a fortunate coincidence the proofs of the paper on "The non-radial oscillations" had arrived the previous day. By working all evening of the 19th and the morning of the 20th, I was able to read and check the proofs; and that was also ready to be returned to the Royal Society on December 20. Also, when I called Wainwright to say that I was sending him the corrected proofs, he informed me that the paper on "The non-radial oscillations of a slowly rotating star" had been passed for publication and that he will try to have the paper included in the March 1991 issue of the Proceedings (i.e. in the issue following the one in which our first paper on the non-radial oscillations will appear).

The fact that on the same day, I would be sending back the proofs of one paper, learn of the acceptance of another, and submit a third, reminded me of my waiting anxiously for the acceptance of my first paper to the Royal Society 62 years earlier.

January 8, 1991

Postscript

The story did not end as happily as I wrote on January 8: a fresh calculation had to be made to the paper on the Lense–Thirring effect; and some sections from Paper II on further amplifications had to be deleted because of a basic misunderstanding of the premises underlying the paper on the flux integral. But Paper III, written subsequently, had a happier ending. Let me take these in order.

1. As already explained on pages 14–16, some questions by John Friedman and related misunderstandings by Bernard Schutz had to be clarified — not altogether to their or my own satisfaction. However, it soon became clear what the source of my own dissatisfaction was: it was simply that I had not addressed myself to the question as to the manner of excitation by a slow rotation of the axial oscillations when the star is already oscillating in its quasi-normal polar mode. Once the problem was formulated with this clarity, the solution was not far to seek. I wrote to Valeria on January 16 formulating clearly the further calculations that had to be done; and she was able to fax me the solution on January 19. It was fortunate that a further note added to this paper on January 21 could be included in the paper, avoiding the writing of a separate paper.

What was surprising to me was that it took considerable effort to convince both Friedman and Schutz that their own earlier uneasiness with regard to the paper was indeed clarified by this last addition.[n]

[n] Eventually, Schultz must have been satisfied since he wrote a nice account of the paper in *Physics World* (issue of August 1991).

2. I had recommended to the Royal Society that Paper II on 'Further Amplifications' be sent to Bernard Schutz as a referee. He took interminably long over his reviewing; and when I called him on the telephone concerning the delay, he explained what the source of his misgiving was. I did not follow his arguments well enough; but on further thought, I realized that our application of the flux integral to describe the flow of gravitational radiation through the star was invalid: it could not be applied as it was derived for quasi-normal modes when σ is complex. I decided to delete the relevant sections to avoid further discussions. After this deletion, the paper had smooth sailing.

3. Discussions with Bob Wald during the early summer of 1990 convinced me that one should examine the axial modes of oscillation for compact stars with radii less than three times the Schwarzschild radius. And during my September 4–11 visit to Rome, I had suggested to Valeria that she examine this problem.

On 5th March Valeria sent some preliminary calculations on the quasi-normal modes of stars with radii between 1.2 and 1.3 Schwarzschild radii. Her comparison with the quasi-normal modes of a black hole was clearly invalid. As I wrote in a long letter on March 5 (and again on March 6), the correct interpretation of her results was different; and I explained in detail my own interpretation (as given in the published paper) with additional suggestions as to what she should do next. She was not sufficiently clear about all the ramifications of the results, and she volunteered to come to Chicago during the fortnight March 25–April 7 so that the paper could be completed and written. I was not certain to what extent I needed to go along with this new project. But when she arrived it became clear that additional calculations were needed, new diagrams had to be made, and that the paper could not be written by her alone — she was quite clear about the last. During the two weeks of her stay, it was indeed possible to have a fully revised $n - 1$ draft of the paper two or three days before her departure. We had to wait for the Graphic Arts Department to make the illustration and the manuscript to be retyped. But somehow it was in fact possible for me to give the final typescript of the paper to her before she set out to the airport.

The paper was communicated to the Royal Society on April 8. The paper did have a happy ending in that Bernard Schutz (with his usual delay) did write on May 28, "I have read your later paper (*Proc. Roy. Soc.* did not send it to me to referee) and like it very much. You are right that it is straightforward, but nevertheless very interesting." (!)

And the Royal Society was prompt enough to have this Paper III published ahead of Paper II.

During the two weeks Valeria was here, we had discussions with Rafael Sorkin, who was as convinced as he was a year earlier, that with the Einstein pseudo-tensor we ought to get the correct flux integral (which we had failed to obtain with the Landau–Lifshitz pseudo-tensor). And Sorkin did notice that, in our earlier calculations with the Einstein pseudo-tensor, we had made an error of sign in its definition. This error in signs invalidated our earlier result since the Einstein pseudo-tensor is not symmetric. I was still not convinced. However, in a discussion between Sorkin, Wald and Valeria, I set out my own views on the matter and also my misgivings. Eventually Sorkin and Wald restated their views in conformity with my own with sufficient clarity that I could pinpoint my own misunderstanding. What was left to do was to go back to first principles and show that the use of the Einstein pseudo-tensor did give the correct final result! But the matter was left in abeyance when Valeria left. However, a week after she left, it was possible to demonstrate that Sorkin was indeed right with respect to the Einstein pseudo-tensor for the vacuum. At the same time, the usual procedure of replacing in the pseudo-tensor, $\mathcal{G}^i_j$, by the energy momentum tensor, T^i_j, simply did not work for the Einstein–Maxwell spacetime. The matter had to be left at this point, since we had to leave for Oxford for my five weeks of lecturing on the Principia. (About these lectures at a later time!)

On my return from Oxford, Sorkin was able to tell me what the term is that must be added to the vacuum Einstein pseudo-tensor when there was a prevalent electromagnetic field. It was not difficult to show that with Sorkin's additional term, one does indeed obtain the correct flux integral, though I was stymied for a while by ignoring the need to be careful with the use of the initial-value equation.

Our present intention was that Sorkin would write a paper explaining his approach to the entire problem of the pseudo-tensor and the conservation of energy on a linearized theory; that we would write a paper detailing our application of the Einstein pseudo-tensor to follow Sorkin's; and that both papers could be communicated to the Royal Society during Valeria's visit in July–August.

Before leaving for Erice, I wrote out an $(n-1)$-draft of the paper with an introduction quoting the early reactions of Weyl, Pauli and Eddington to the Einstein pseudo-tensor.

June 1991

Let me continue with the story of the Einstein pseudo-tensor. As I wrote earlier, our intention was that Sorkin should write an account of his own on the origins and the meaning of pseudo-tensors and their relation with Noether currents. He completed a first draft of his paper by the end of July. He then scrutinized the $n-1$ draft that I had written. He pointed out that what I had written about the suitability of the Einstein pseudo-tensor for the purposes of evaluating the flux integral was not valid and that the correct way of looking at the problem was in terms of the identity,

$$\mathcal{E}^i_{0,i} = \frac{1}{2}\, \mathcal{G}^{ij}\frac{\partial g_{ij}}{\partial t}\,,$$

and its first and second variations. This identity clarified for me for the first time the problem that had been troubling me for some three years. Sorkin's suggestion required the rewriting of the earlier sections of the paper. Both papers were communicated to the Royal Society a week before Valeria left. (As a postscript, I may add that both papers were accepted by the Royal Society within a week of their submission.)

While the paper on pseudo-tensors was still pending, I had discussions with Roland Winston, in which I expressed my uneasiness with the relationship between the manner in which we had solved the problem of resonant scattering by stars and the conventional treatment of the Breit–Wigner formula in quantum mechanics. Some days later, Roland called to say that perhaps the theory of the Regge poles could be used to clarify my doubts. While Roland's idea was not entirely to the point, I soon became convinced that the Regge theory provided an alternative approach to the problem with important ramifications for the use of the flux integral we had derived. I made a preliminary outline of my own development and showed it to Winston. He was enthusiastic about this new approach; but the matter had to be tested. The final court of appeals — in all such cases is numerical confirmation.

I called Valeria before her departure from Rome, indicating that an alternative approach to our problem via Regge's theory had opened up and asking her to bring with her programs relating to our earlier calculations. By the time she arrived, I had essentially completed the application of Regge's theory to the resonant scattering of axial gravitational waves. She started on this work with her usual enthusiasm and efficiently, and within a week of her arrival, the application of the theory of Regge poles to the axial modes of oscillation was completed; and the basic ideas were confirmed by the numerical work.

Turning next to the polar oscillations, it was not too difficult to isolate the Regge poles that corresponded to the imaginary part of the complex frequency belonging to the quasi-normal modes. The resolution of the problem relating to the growth of gravitational energy through the star was not as straightforward. The problem was a conceptual one. After many trials, the solution of the problem became transparent once we realized that all that was required was the analytic continuation of the flux integral (zero on the real axis) into the complex ℓ-plane. The necessary calculations were completed before Valeria left.

After Valeria left, I had to concentrate on the opening address I was to give at Erice on Daniel Chalonge. And there was also the unexpected interruption to go to Boulder to receive the Abott award. (I am sure that Roland Winston was behind it, although he disclaims any responsibility!) Besides, I also wanted to have the $n-1$ draft of the Regge paper written before we went to Erice and later to Rome.

The three days in Erice were largely memorable because of our meeting with Karen Chalonge after some 45 years. It was an unexpected experience to find the friendship of long ago had retained its freshness.

The week in Rome was mostly spent in visiting the Sistine Chapel, the Vatican Museum, the Coliseum, and various other sites. Valeria was very generous in spending almost the entire week with us taking us around in her car. But we did find the time to review the $n-1$ draft of the Regge paper that I had brought with me. She made a number of suggestions for improving the manuscript in several places and thought that a concluding section stating the premises of our paper clearly was essential.

On returning to Chicago, I revised the paper and wrote a fresh $n-1$ draft. As I wrote the concluding section comparing the approach to the scattering theory as developed in our series of papers with the conventional approaches I felt that I might be reading on the toes of the physicists. I therefore asked Peter Freund to scrutinize the paper with maximum severity. Peter did read the section carefully, and pointed out a misunderstanding on my part on quantum-theory side and clarified the basic issues that were involved. The final version of the concluding section owes a great deal to Peter's criticisms.

All the revisions took a fair amount of time. The final manuscript was typed and communicated to the Royal Society on September 18.

I forgot to write about the various illustrations which form an essential part of the paper. It was not too easy to organize and arrange the many illustrations. It required more time than I had originally thought. The excellent cooperation that I received from the Graphic Arts department — and especially from Edward Poole — was very essential.

I felt that I had now a fairly complete understanding of the problem of the non-radial oscillations of stars. I did not have any unanswered questions in my mind; and I had no plans for further work.

October 1991

Continued Efforts IV (November 1991–December 1994)

At long last, I am returning to my narration of my 'Continued Efforts' after a lapse of three years during which time my book on Newton's Principia was begun and completed. It was also the period of my failed efforts to complete the series on 'Non-radial oscillations' by solving the last remaining problem. But to return to October 1991.

In some ways, 1991 was a successful year. The papers on the resonant axial modes of oscillations, the application of the Regge Theory of scattering to determine the flow of gravitational radiation through the star, and the resurrection of the Einstein pseudo-tensor (with the help of Rafael Sorkin) were all written after the Oxford Lectures (April 17–May 22) and the contract with the Clarendon Press for my book on the Principia had been negotiated.

And that was not all! I had to prepare and give a brief talk on Abbot on the occasion of the Abbot Award by the American Solar Energy and the opening address at the Chalonge-Symposium at Erice. (Both these addresses gave me the opportunity to return to my early interests in the continuous spectrum of the Sun and in H^-.) Besides, I had to give five Russell Marker Lectures at

the Pennsylvania State University. Finally, we had to go to Crete
for the presentation of the first Basilis Xanthopoulos Award to
Christadoulu.

The year 1992 began with the Mathematics Colloquium (at the
invitation of Raghavan Narasimhan) on 'Some Propositions from
Newton's Principia.'

The simultaneous arrival of Persides° heralded that my book on
the Principia cannot be postponed any further. Indeed during 1992
I did write 13 Chapters. But my efforts with Valeria, while dormant,
did not cease altogether.

Contrary to what I wrote in October 1991, I felt that the
series on the 'Non-radial oscillations' would not be complete without
showing the exact emergence of the Newtonian limit. We attempted
to solve this problem during Valeria's visits in July 26–August 29
and October 18–November 11 in 1992.

Our approach to the problem at this time was to separate
the curvature terms in $\nu_{,r}$ and $\mu_{2,r}$ in the basic Equations (48)–
(51) (in S. C. and V. F., *Proc. Roy. Soc. A* **432** (1991)) from the
rest and consider the scattering of free gravitational waves by the
Newtonian star in the manner of the Rayleigh–Mie theory of the
scattering of electromagnetic waves by spherical particles. We found
out after many trials that we were pursuing a false trail. But we did
salvage our treatment of the propagation of free gravitational waves
in Minkowskian space; and we sent a short paper on this to the Royal
Society during Valeria's visit early in 1993 (February 21–March 1).
While this was a small accomplishment for the time and effort spent,
the identity (A4) derived in this paper was to lead to the eventual
solution of the problem. A sporadic effort during Valeria's second
visit in August 1–14 was equally futile.

°At the time, when Basilis was assassinated, Persides, as a result of an act of bravery
in attempting to prevent the assassin from further acts of violence, received deadly
injuries. I thought that a change of scene and interests might accelerate his recovery.
I invited him to come to Chicago for a few months and assist me in critically reading
my manuscript as it progressed. By his efforts during the following two years, Persides
was an enormous and an essential help.

There was one other matter which I kept on ruminating in spite of the continuing pressure of the Principia. And that was on the question that arose during the memorable conversation I had with Penrose during our brief visit to Oxford in October 1989 on our return from our vacation in Vienna (see page 219 of Continued Efforts III). It will be recalled that during that conversation, the similarity of Monet's motivations in painting his 'Series Paintings' and my own motivations in my series of papers on black holes, colliding waves, and scattering of gravitational radiation by stars became apparent. After our return from Oxford, I studied the various catalogues of Monet's paintings particularly "Monet in the 90's: The Series Paintings" issued by the Art Institute of Chicago on the occasion of the exhibition of Monet's series paintings in 1989). The little booklet "Monet by Monet" in the series 'Artists by Themselves" was specially useful. In this context, I should mention the handsome collection of Monet's Series Painting that Jack Cella and Ipsita Chatterjee assembled and presented to me.

Eventually, I wrote an essay on "The Series Paintings of Claude Monet and the Landscape of General Relativity". My intention was to include it among the articles that I had solicited for a theme issue on "Classical general relativity" for the *Transactions of the Royal Society*. But the number of pages allowed for the issue did not allow for its inclusion.

My first attempt to present the essay as a lecture at Rutger's University was a failure. My second attempt to give it as my Dedication Address at the inauguration of the Inter University Centre of Astronomy and Astrophysics at Pune in December, 1992 was an even worse failure. In both cases the arrangements were very bad: e.g. in Pune the lecture had to be given outdoors in darkness! But IUCAA did publish my lecture in a handsome format.

I made a third attempt to give it as my opening address at the 7th Marcel Grosmann meeting at Stanford University on July 24, 1994. I thought it went off moderately well. And lastly, I gave the same lecture at the Physics Department Colloquium in October (1994). That ends this matter.

Among other incidental matters that occurred during 1994, I may mention my presentation of the bust of Ramanujan to the Royal Society in May, 1994. At the dinner arranged by Atiyah on this occasion (also, the 50th anniversary of my election to the Royal Society), Anne Davenport, David and Kate Shoenberg, Roger and Vanessa Penrose, Richard Dalitz and his wife, Meggie and John Weston-Smith, and Valeria were present as my invited guests. Penrose made a generous speech; and my own remarks on Ramanujan's bust is to be published in the *Notes and Records of The Royal Society* (Jan. 1995).

And finally at the 44th Conference of Nobel Laureates in Lindau, I gave a lecture on Newton & Michelangelo (published in *Current Science* **67** (1994)) along with an earlier article "On Reading Newton's Principia at age past eighty" dedicated to Kothari.

1994, November 27

Postscript

In the leisure that followed the completion of my book on the Principia I began to think once again about the problem of a fully relativistic treatment of Newtonian oscillations. I casually noticed a remarkable feature of the four basic equations that we had derived already in our first paper (Equations (72)–(75) in Paper I) — a feature that stares in the face — once noticed! The four equations split into two pairs: a pair that survives in the Minskowian limit when all the terms depending on the curvature of the space-time are ignored — equations which in essence describe the propagation of free gravitational waves in Minkowski space; and the other pair which vanishes identically as each of the terms in this pair is directly dependent on the curvature expressed by $\nu_{,r}, \mu_{2,r}$ or $e^{2\mu_2} - 1$. If one removes the common proportionality factor G/c^2 of these terms, the equations remain finite after ignoring terms that are of second and higher orders in the curvature.

But Valeria's first efforts at calculations had errors. And she concluded that "we must accept" that the method does not work. I was not convinced and I tried a different approach via the identity in the Appendix to the paper on Spherical Gravitational Waves. In this manner, I obtained a pair of coupled second order linear differential equations in two variables which combine in a common scheme both the Minkowskian and the Newtonian limits. The equations were so beautiful that I was convinced that the approach must be right.

Again, Valeria thought that equations must be linearly dependent. Again, I was convinced that she was wrong and I essentially obtained the correct indical equation. But there was an oversight which I corrected later. But all was well that ends well.

Valeria was too busy to do the calculations. But I wrote the paper anyway. The concluding paragraphs state my final views.

The equations derived in the manner described, determine the exact Newtonian characteristic frequencies of non-radial oscillations by allowing two linearly independent singularity-free solutions at the centre and satisfying the boundary conditions that will ensure that no gravitational radiation emerges (even as they do not for dipole oscillations of fully relativistic stars). In other words, the distinguishing characteristic of the scattering of gravitational waves by Newtonian stars is that no gravitational radiation emerges.

Several questions occur in contexts larger than in the example considered: Should one expect, quite generally, that equations describing the scattering of gravitational waves by the space-times of closed Newtonian systems will combine in the same manner in a common scheme both of the Minkowskian and the Newtonian limits? Can one expect that no gravitational radiation will emerge from such Newtonian system as a theorem of general validity?

1995, February 14

1992:

28 February: U of C Mathematics Colloquium "Some propositions from Newton's Principia"

13–14 April: New Burnswick, NJ, Rutgers University, Department of Philosophy — Rutgers Distinguished Lecture. Talk titled "The Series Painting of Claude Monet and the Landscape of General Relativity"

27 April: University of California, Berkeley; The Indo-American Community Chair in India Studies Lecture: "India's Contributions to the Physical Sciences: Before and After Independence".

28 April–1 May: Nobel Laureate Lecture Series, Long Beach, CA., California State University — Discussion with students on "Truth and Beauty".

14–15 May: Syracuse University for "Walifest" Lecture: "Scattering of Gravitational Waves by Stars and by Black Holes".

23–25 December: Oxford: Clarendon Press & Roger Penrose.

27–30 December: Pune, India, IUCAA Dedication Address: "The series Paintings of Claude Monet and the Landscape of General Relativity".

30 December: Madras

Visiting Scholars:

17 February– 31 May: Sotirios Persides

26 July–29 August; 18 October–11 November: Valeria Ferrari

Papers Published:

On the non-radial oscillations of a star. IV. An application of the theory of Regge poles (with Valeria Ferrari), *Proc. Roy. Soc. London A* **437** (1992) 133–149.

1993:

3–18 January: Bangalore

3 March: On spherical free gravitation waves (with Valeria Ferrari), *Proc. Royal Society*, London communicated)

17–18 April: The Lincoln Academy of Illinois, Laureate Convocation

12 May: Northwestern University, Evanston, Illinois: "The Relevance of the Principia for a Student of Today"

28–30 September: Oxford, Clarendon Press: Chapters 1–21 for editing

1–5 October: Amsterdam, International Conference: "Child Labor and Child Abuse"

28 October: The Chicago Academy of Sciences symposium: "Motivations in Science and in the Arts"

6–27 December: India: Delhi, Madras, Bangalore

Visiting Scholar:

21 February–1 March; 1–14 August: Valeria Ferrari

Papers Published:

On spherical free gravitation waves (with Valeria Ferrari), *Proc. Roy. Soc. London A* **443** (1993) 445–449.

The Series Paintings of Claude Monet and the Landscape of General Relativity, *Proc.* (IUCAA, India, 1993).

On the occasion of the Charles Greeley Abbot Award by the American Solar Energy Society, *Solar Energy* **51**, No. 3 (1993) 233–235.

Daniel Chalonge and the problem of the abundance of Hydrogen, *Proceedings of First Course: Current topics in Astro-fundamental Physics* (World Scientific Publishing Co.).

1994:

13–15 January: American Mathematical Society, Cincinnati: "Some Propositions from Newton's Principia".

May 3: USAAPCC Excel 2000 Award, Washington, DC.

9–11 May: The Royal Society of London, to present the bust of Ramanujan as a gift to the Royal Society and to attend the Ascension Feast at Trinity.

21 June–1 July: The 44th Conference of Nobel Laureates: Invited attendance and to give a lecture on "Newton and Michelangelo".

24–30 July: Seventh Marcel Grosmann meeting at Stanford University: "The Series Paintings of Claude Monet and the Landscape of General Relativity".

13–24 December: India

Visiting Scholar:

8–20 August: Valeria Ferrari

Papers Published:

On reading Newton's Principia at age past eighty, *Current Science* **67** (1994) 495–496.

Newton and Michelangelo, *Current Science* **67** (1994) 497–499.

NOTES & COMMENTS
Kameshwar C. Wali

By no means exhaustive, these notes are meant to provide to the interested reader quick references to the papers and personalities mentioned in the text as they appear in the original text. Names with asterisks and the year refer to Chandra's students and the year they got their degrees.

I. A History of Papers on "Radiative Equilibrium" (1943–1947)

1. Placzek [George (1905–1955)], a Czech physicist, born in Brno, Moravia, one of the important physicists of the 20th century, who made seminal contributions to the fields of molecular physics, scattering of light from liquids and gases, the theory of atomic nucleus and the interaction of neutrons with condensed matter. Chandra met him in 1932 in Copenhagen, at Niels Bohr Institute and had become friends.

2. L. Gratton, *Soc. Astr. Italiano* **10** (1937) 309.

3. Aberdeen: the Ballistic Research Laboratory at the Aberdeen Proving Grounds (APG), where Chandra was involved in war related work from 27 January 1943 till the end of the war. He commuted between Yerkes and APG — three weeks at Yerkes, three weeks at APG. He was elected to the Royal Society of London in 1944.

4. Henyey [Louis George] was an astronomer at Yerkes, who had earned his doctorate from the University of Chicago with a mathematical thesis on the topic of reflection nebulae. Chandra's reputation as a teacher, and his youth and enthusiasm for research had attracted students for all parts of the world. Sahade [Jorge], Cesco [Carlos] and Krogdahl [Margaret Kiess] were among them.

5. Hopf–Bronstein relation refers to solutions of a class of Integro-differential equations studied by mathematicians, E. Hopf, J. Bernstein and N. Wiener. For more details and for an exhaustive review of the whole subject, see *The Transfer of Radiation in Stellar Atmospheres*, S. Chandrasekhar, *Bulletin of the American Mathematical Society* **53**, no. 7 (1947) 641–711.

6. Non-gray atmosphere. A stellar-atmosphere in local thermo-dynamic equilibrium with a constant absorption coefficient is described as "gray" atmosphere. If the absorption coefficient is a function of frequency of radiation at each point, the corresponding stellar atmosphere is described as non-gray. Münch [Guido* (1946)] to be one of the graduate students. Ref. to Unsöld: A. Unsöld, *Physik der Sternatmospharen* (Springer, Berlin, 1938), pp. 113–116.

7. Van der Mondie [Vandermonde] determinant is the determinant of a matrix with terms of a geometric progression in each row. Kopal [Zdeněk], a Czech astronomer, who was at the time at Harvard College Observatory. Ambartsumian: Chandra had met Ambartsumian [Victor Amazaspovitz] in the summer of 1934 in Leningrad. Because of the Second World War, he first became aware of his paper on principles of invariance only in the summer of 1945 (C.R. (*Doklady*) *Acad. URSS* **38** (1943) 257.

8. Kuiper [Gerard], the observational astronomer had joined Yerkes at the same time as Chandra in 1937.

9. A. Schuster, *M.N.* **40** (1879) 35, M. Minnaert, *Zs. f. Ap.* **1** (1930) 209, H. Zanstra, *M.N.* **101** (1941) 250. Herzberg [Gerhard], a pioneering physicist, physical chemist and spectroscopist, Struve [Otto] was the Director of Yerkes Observatory.

10. Teller [Edward], Breit [Gregory], von Neumann [John].

11. Stokes [George Gabriel] parameters that describe the polarization states of electromagnetic radiation. On the composition and resolution of streams of polarized light from different sources, *Trans. Camb. Phil. Soc.* **9** (1852) 399.

12. Lindblad [Bertil], Director of Stockhom Observatory, Krishnan [K. S.], Raman's collaborator in the discovery of Raman Effect. Hamilton [D. R.], related work concerning resonance scattering, *Astrophys. J.* **106** (1947) 457.

13. Ledoux [Paul] was a graduate student. Titchmarsh [Edward Charles] well-known British Mathematician at Oxford. Author of the well known textbook, *Theory of Functions*.

14. Davenport [Harold], a pure number theorist and a mathematician. Chandra had known Davenport since his Cambridge days.

15. Bengt: Strömgren [Bengt], Frances: [Miss Frances Herman].

16. van de Hulst [H. C. van de Hulst], a Dutch astronomer, who had just completed his doctoral thesis and was a post-doc at that time at Yerkes.

II. Turbulence; Hydromagnetism (1948–1960)

1. Karl Schwarzschild [1873–1916], the well known German physicist, noted for Schwarzschild solution in Einstein's equations and many other important contributions. Eddington [Arthur Stanley (1882–1944)]. Jeans [James Hopwood (1877–1946)]. Milne [Edward Arthur (1896–1950)].

2. Wigner [Eugene (1902–1995)]. An example of Chandra's pattern of beginning research in a new area, give seminars, teach a course or invite an expert to teach him. The paper by Kármán [T. von] and Howarth [L.] probably refers to the equation known as Kármán–Howarth equation derived from Navier–Stokes equations in the case of isotropic turbulence. On the statistical theory of isotropic turbulence, *Proc. Roy. Soc. Ser. A* **164** (1938) 192–215. Taylor [G. I.], *Proc. R. Soc. London A* **164** (1938) 476. Papers by Kolmogoroff [Andrey Nikolaevich Kolmogorov], *C.R. Acad. Sci. U.S.S.R.* **30** (1941), 301; **32** (1942) 16 and Batchelor [G. K.], *Proc. Cambridge Phil. Soc.* **43** (1947) 533. Heisenberg [Werner], *Z. Phys. Rev.* **124** (1948a) 628 and *Proc. Roy. Soc. A* **195** (1948b) 402. The audience at the evening seminars included graduate students Münch [Guido], Osterbrock [Donald E.* (1952)], Code [Arthur D.* (1950)]. For the official course on Radiative Transfer, two that registered were Yang [Chen Ning] and Lee [Tsung-Dao], Donna [Elbert D.].

3. Robertson [H. P.] The invariant theory of isotropic turbulence, *Proc. Cambridge Philos. Soc.* **36** (1940) 209–223. André Weil [1906–1998], a mathematician and Gregor Wentzel [1898–1978], a theoretical physicist, a colleague at the University of Chicago. Uhlenbeck [George Eugene (1900–1988)], a theoretical physicist known among other, the hypothesis of electron spin along with Goudsmit [Samuel Abraham (1902–1978)].

4. Spitzer [Lyman (1914–1997], Schwarzschild [Martin (1912–1997)] renowned astrophysicists. Lord Adrian [Edgar (1889–1977)], Master of Trinity College, winner of Nobel Prize for Physiology and Medicine in 1932.

5. Jeans' Criterion: about gravitational stability of an infinite homogeneous (non-turbulent) medium, J. H. Jeans, *Astronomy and Cosmology* (Cambridge University Press, London, 1920), pp. 345–349.

6. Rayleigh: [Lord Rayleigh, Strut John William (1842–1919)], Jeffreys [Sir Harold (1891–1989)] Fermi [Enrico (1901–1954], Burbidge [Geoffrey (1925–2010)] Couette [Maurice Marie Alfred, a Professor of Physics at the French University of Angers in the late 19th century] Couette flow refers to the laminar flow of viscous fluid between two parallel plates, one of which is moving relative to the other. Taylor-Coutte flow refers to the flow between two co-axial cylinders. Morgan [William W.].

7. Hylleraas [Egil], *Zs. F. Phys.* **60** (1930) 624. Limber [Nelson D* (1953)], vector **H** and vector Ω refer to the magnetic field and angular velocity. The paper with Limber, *The Astrophysical Journal* **118**, no. 1 (1954) 10–13.

8. Raymond. Hide, The character of the equilibrium of an incompressible heavy viscous fluid of variable density: an approximate theory, *Proc. Cambridge Philos. Soc.* **51** (1955) 179–201.

9. Fultz, D. and Nakagawa, Y., *Proc. Roy. Soc. A* **231** (1955) 211.

10. Dyson [Freeman], Goldberger [Marvin].

11. [Going to Madison to consult with Wigner. A habit Chandra cultivated as we shall discuss later, flying to Oxford, England to consult with Roger Penrose]. Prendergast [Kevin H.] Paper with Prendergast, *Proceedings of the National Academy of Sciences* **42**, no. 1 (1956) 5–9. Backus [George E.* (1956)], paper with Backus, *ibid.* **42**, no. 3 (1956) 105–9. Heisenberg [Werner (1901–1976)].

12. P. 16. It is not clear what paper Chandra is referring to in the last paragraph on this page.

13. Trehan [Surinder K.* (1958)], Siciy [...].

14. Allison [...], setting up of hydromagnetic laboratory. ONR [Office of Naval Research?].

15. Rosenbluth [Marshall (1927–2003)], a noted Plasma physicist. Paper of Lüst and Schlüter, *Kraftfreie Magnetfelder, Z. f. Astrophysik* **34** (1954) 263–82. Ken Watson [K. M.] and Murph Goldberger [M. L.] Well-known theoretical physicists, known for their work on Scattering Theory. S. Chandrasekhar, On force free magnetic fields, *Proc. Nat. Acad. Sci.* **42** (1956) 1–5.

16. Metropolis [...], Sykes [John].

17. Onsager [Lars (1903–1976)] Donnelly [R. J.] was at the Institute for the study of metals at the University of Chicago. The two papers regarding Helium II between rotating cylinders: The Hydrodynamic Stability of Helium II Between Rotating Cylinders, I & II, *Proceedings of the Royal Society A* **241** (1957) 9–28 and 29–36, Paper I is with Donnelly. The Rumford Medal Lecture: Thermal Convection, *Proceedings of the American Academy of Arts and Sciences* **86**, no. 4 (1957) 323–39.

18. Mathematica Paper: The stability of viscous flow between rotating cylinders. *Mathematica* **1**, 5–13. The joint paper with Reid [W. H.]: On the expansion of functions which satisfy four boundary conditions, *Proceedings of the National Academy of Sciences* **43**, 521–27.

19. Papers published in the Annals of Physics: Properties of ionized gas of low density in a magnetic field, III (with A. Kaufman and K. M. Watson), *Annals of Physics* **2**, 435–70; and *ibid.* IV, *Annals of Physics* **5**, 1–25.

20. The stability of viscous flow between rotating cylinders in the presence of a magnetic field, II (with D. Elbert), *Proceedings of the Royal Society A* **262**, 443–54. R. J. Donnelly and M. Ozima, Hydromagnetic stability of flow between rotating cylinders, *Phys. Rev. Lett.* **4** (1960) 497–8.

21. Edmonds [Frank N. Jr.* (1950)] The oscillations of a viscous liquid globe, *Proceedings of the London Mathematical Society* **9**, 141–49. On the continuous absorption coefficient of the negative hydrogen ion, V (with D. Elbert), *The Astrophysical Journal* **128**, 633–35.

22. The thermodynamics of thermal instability in liquids, in *Max-Planck-Festschrift* (Veb Deutscher Verlag der Wissenschaften, Berlin, 1958), pp. 103–14.

23. The stability of inviscid flow between rotating cylinders, *Journal of the Indian Mathematical Society* **24**, 211–21. Friedman [John L.* (1973)].

24. Vandervoort [Peter O.* (1960)]. Boussinesq approximation in deriving the hydrodynamical equations: J. Boussinesq, *Théorie Analytique de la Chaleur* **2**, 172, Gauthier-Villars, Paris, 1903.

25. Taylor–Proudman Theorem: All steady slow motions in a rotating inviscid fluid are necessarily two-dimensional. G. I. Taylor, Experiments with rotating fluids, *Proc. Roy. Soc. (London) A* **100** (1921) 114–21; J. Proudman, On the motion of solids in a liquid possessing vorticity, *Proc. Roy. Soc. (London) A* **92** (1916) 408–22.

26. E. Lyttkens, The onset of convection in a mantle of sphere with a heavy core (unpublished). Di Prima: R. C. Di Prima, The stability of viscous flow between rotating concentric cylinders with a pressure gradient acting round the cylinders, *J. Fluid Mech.* **6** (1959) 462–8. Short paper: The hydrodynamic stability of inviscid flow between coaxial cylinders, *Proceedings of the National Academy of Sciences* **46**, 137–41.

27. Reid: W. H. Reid, The stability of non-dissipative Coutte flow in the presence of an axial magnetic field, *Proc. Nat. Acad. Sci.* **46** (1960) 370–3. Niblett: E. R. Niblett, The stability of Coette flow in an axial magnetic field, *Canadian J. Phys.* **36** (1928) 1509–25. D. L. Harris and W. H. Reid, On orthogonal functions which satisfy four boundary conditions. I. Tables for use with Fourier-type expansions, *Astrophys. J. Supp., Ser.* **3** (1958) 448–52. Dalitz [Richard Henry (1925–2006)]. A colleague, noted for sid seminal contributions in Particle physics [Dalitz Plot, tau-theta puzzle leading to the discovery of parity violation].

28. Kelvin–Helmholtz instability: The instability that arises when two superposed fluids flow one over the other with a relative horizontal velocity; the instability of the plane interface between the two fluids, when it occurs is called Kelvin–Helmholtz in stability.

29. Limber [D. Nelson* (1953)].

30. The paper he refers to his probably that of L. B. Bernstein, E. A. Freiman, M. D. Kruskal and R. M. Kulsrud, An energy principle for hydromagnetic stability problems, *Proc. Roy. Soc. A* **244** (1958) 17–40.

31. Lebovitz [Norman* (1961)].

III. The Development of the Virial Method and Ellipsoidal Figures of Equilibrium (1960–1970)

1. On the pulsation of a star in which there is a prevalent magnetic field (with D. N. Limber), *The Astrophysical Journal* **119**, 10–13; Lord Rayleigh, On a theorem analogous to the virial theorem, *Scientific Papers*, iv. 491–3, Cambridge, England, 1903; E. N. Parker, Tensor virial equations, *Phys. Rev.*

96 (1954) 1686–9; Problems with gravitational stability in the presence of a magnetic field (with E. Fermi), *Astrophys. J.* **118** (1953) 116–41; P. Ledoux, Stellar stability, *Handbuch der Physik* **51** (1958) 605–88.

2. A theorem on rotating polytropes, *The Astrophysical Journal* **134** (1961) 662–64.

3. On super-potentials in the theory of Newtonian gravitation (with N. Lebovitz), *The Astrophysical Journal* **135** (1962) 238–47; On the oscillations and stability of rotating gaseous masses, *The Astrophysical Journal* **135** (1962) 248–60; An interpretation of double periods in β Canis Majoris stars, *The Astrophysical Journal* **135** (1962) 305–6; Nehru [Jawaharlal (1889–1964); Prime Minister of India (August 15, 1947 till May 27, 1964)].

4. E. J. Routh, *A Treatise on Analytical Statics* **2** (Cambridge, England, Cambridge University Press, 1892), **67**, 194–231, 239–54. N. M. Ferrers, On the potentials of ellipsoids, ellipsoidal cells, elliptic laminae, and elliptic rings, of variable densities, *Quart. J. Pure and Appl. Math.* **14** (1877) 1–22; The potentials and the superpotentials of homogeneous ellipsoids (with N. Lebovitz), *The Astrophysical Journal* **136** (1962) 1037–47; On superpotentials in the theory of Newtonian gravitation II. Tensors of higher rank (with N. Lebovitz), *The Astrophysical Journal* **136** (1962) 1032–36.

5. On the point of bifurcation along the sequence of the Jacobi ellipsoids (with N. Lebovitz), *The Astrophysical Journal* **136**, 1048–68; The points of bifurcation along the Maclaurin, the Jacobi and the Jean sequences, *The Astrophysical Journal* **137**, 1185–1202. On the occurrence of multiple frequencies and beats in the β Canis Majoris stars (with N. Lebovitz), *The Astrophysical Journal* **136** (1962) 1105–7.

6. On the oscillations and the stability of rotating gaseous masses, II. The homogeneous, compressible model (with N. Lebovitz), *The Astrophysical Journal* **136** (1962) 1069–81; An approach to the theory of equilibrium and the stability of rotating masses via the virial theorem and its extensions, in *Proc. Fourth U.S. National Congress on Applied Mathematics* (1962), pp. 9–14.

7. Lyttleton [Raymond [1911–1995)] An eminent British astronomer, Lynden–Bell [Donald], an eminent English astrophysicist. On the stability of the Jacobi ellipsoids (with N. Lebovitz), *The Astrophysical Journal* **137** (1963) 1142–61; On the oscillations of the Maclaurin spheroid belonging to the third harmonics (with N. Lebovitz), *The Astrophysical Journal* **137**, 1162–71; The equilibrium and the stability of the Jeans spheroids (with N. Lebovitz), *The Astrophysical Journal* **137** (1963) 1172–81. Silliman Lectures [The Silliman Foundation Lectures; the foundation established in memory of Mrs. Hepsa Ely Silliman, the President and Fellows Of Yale University].

8. The case for astronomy, *Proc. American Philosophical Society* **108** (1964)

1–6; The ellipticity of a slowly rotating configuration (with P. Roberts), *The Astrophysical Journal* **138** (1963) 801–8.

9. Roche limit: Limit on the angular velocity of an infinitesimal satellite rotating about a rigid spherical planet in a circular Keplerian orbit. A general variational principle governing the radial and the non-radial oscillations of gaseous masses, *The Astrophysical Journal* **138** (1963) 896–97 and **139** (1964) 664–74. Ostriker* [Jeremiah P. (1964)]. Clement* [Maurice J. (1965)].

10. Darwin [George Howard (1845–1912)] Astronomer and Mathematician. On the figures and stability of a liquid satellite, *Phil. Trans. R. Soc. (London)* **206** (1906); *Scientific Papers* **3** (Cambridge, England, Cambridge University Press, 1910), 436; The equilibrium and the stability of the Darwin ellipsoids, *The Astrophysical Journal* **140** (1964) 599–620.

11. Dynamical instability of gaseous masses approaching the Schwarzschild limit in general relativity, *Physical Review Letters* **12** (1964) 114–16; erratum, *Physical Review Letters* **12** (1964) 437–8; The dynamical instability of the White-Dwarf configurations approaching the limiting mass (with R. F. Trooper), *The Astrophysical Journal* **140** (1964) 417–33; Dynamical instability of gaseous masses approaching the Schwarzschild limit in general relativity, *The Astrophysical Journal* **140** (1964) 417–33; Uhlenbeck [George Eugene (1900–1988)] Dutch born American theoretical physicist noted for postulating electron spin long with Samuel Goudsmit.

12. The equilibrium and the stability of the Dedekind ellipsoids, *The Astrophysical Journal* **141** (1965) 1043–55; The stability of a rotating liquid drop, *Proc. Royal Society A* **286** (1965) 1–26; The equilibrium and the stability of the Riemann ellipsoids, *The Astrophysical Journal* **142** (1965) 890–921.

13. Basset A. B., *A Treatise on Hydrodynamics* (Cambridge, Eng.: Deighton Bell & Co.; reprinted 1961, Dover Publications, New York); The equilibrium and the stability of the Riemann ellipsoids, II, *The Astrophysical Journal* **145** (1966) 842–77.

14. Ellipsoidal figures of equilibrium — An historical account, *Communications on Pure and Applied Mathematics* **20** (1967) 251–65; The virial equations of the fourth order, *The Astrophysical Journal* **152** (1968) 293–304; The pulsations and the dynamical stability of gaseous masses in uniform rotation (with N. R. Lebovitz), *The Astrophysical Journal* **152** (1968) 267–91.

15. Lee [Edward*(1968)].

16. A tensor virial-equation for Stellar dynamics (with E. P. Lee), *Monthly Notices of the Royal Astronomical Society* **139** (1968) 135–39; The effect of viscous dissipation on the stability of the Roche ellipsoids, *Publications of the Ramanujan Institute* **1** (1969) 213–22.

17. Jawaharlal Nehru Memorial Lecture: Astronomy in Science and Human Culture, *Indraprastha Press*, 22 pp.

18. Interesting postscript about the last section and work to complete it, Clement [Maurice J.* (1965)].

19. The instability of the congruent Darwin ellipsoids, *The Astrophysical Journal* **157** (1969) 1419–1434; The instability of the congruent Darwin ellipsoids, II, *The Astrophysical Journal* **160** (1970) 1043–48; The book based on the Silliman Lectures, *Ellipsoidal Figures of Equilibrium* (New Haven and London, Yale University Press, 1969) turned out to be a monumental monograph.

IV. General Relativity (1962–1969)

1. The Mathematical Theory of Relativity by A. S. Eddington [First published in 1923; Second edition reprinted in 1930]. McCrea [William H. (1904–1999)], a British astronomer and mathematician Chandra met during his Cambridge days when attending Royal Society meetings in London. They became close friends. Milne [Edward A.] *Kinematic Relativity; A Sequel to Relativity, Gravitation and World Structure* (Oxford Clarendon Press, 1948). Gregor [Gregor Wentzel].

2. Schrödinger [Erwin (1887–1961)] *Space-Time Structure* (Cambridge University Press, 1950). Weatherburn [Charles E. (1884–1974)], an Australian mathematician. Tolman [Richard C. (1881–1948)], American theoretical physicist and physical chemist, noted also for his contributions to Relativity and Cosmology. *Relativity, Thermodynamics and Cosmology* (Oxford Clarendon Press, 1934). The Geodesics in Gödel's Universe (with J. P. Wright* (1961)), *Proceedings of the National Academy of Sciences* **48** (1961) 341–47.

3. A. Einstein, L. Infeld and B. Kaufman, *Ann. Math.* **39** (1938). The virial theorem in general relativity in the post-Newtonian approximation (with G. Contopoulos), *Proc. National Academy of Sciences* **49** (1963) 608–13. Misner [Charles W.] and Zapolsky [Harold], Relativists at Rutgers University.

4. Post-Newtonian equations of hydrodynamics and the stability of gaseous masses in general relativity, *Physical Review Letters* **14** (1965) 241–44. The post-Newtonian equations of hydrodynamics in general relativity, *The Astrophysical Journal* **142** (1965) 1488–1512. The post-Newtonian effects of general relativity on the equilibrium of uniformly rotating bodies, I. The Maclaurin spheroids and the virial theorem, *The Astrophysical Journal* **142**, 1513–18. The stability of gaseous masses for radial and non-radial oscillations in the post-Newtonian approximation of general relativity, *The Astrophysical Journal* **142**, 1519–1540.

5. Henk [H. C. van de Hulst]. Spiegel [Edward A.] Astronomy Department, Columbia University Cowling [Thomas Gorge (1906–1990)], an English mathematician and astronomer. Synge [John Lighton (1897–1995)], an

Irish mathematician and physicist. Fokker [Adriaan (1887–1972], a Dutch physicist. Fock [Vladimir (1898–1974)], a Soviet physicist.

6. The post-Newtonian effects of general relativity on the equilibrium of uniformly rotating bodies, II. The deformed figures of Maclaurin spheroids, *The Astrophysical Journal* **147** (1967) 334–52. The post-Newtonian effects of general relativity on the equilibrium of uniformly rotating bodies, III, *The Astrophysical Journal* **148** (1967) 621–44. On a post-Galilean transformation appropriate to the post-Newtonian theory of Einstein (with G. Contopoulos), *Proceedings of the Royal Society A* **298** (1967) 123–41.

7. Trautman [Andrez], a Polish physicist specializing in Relativity. Penrose [Roger], Rouse Ball Professor of Mathematics at Oxford. Nutku [Yavuz* (1969)].

8. Greenberg [Philip J.* (1971)] Landau–Lifshitz energy-momentum pseudotensor; See L. D. Landau and E. M. Lifshitz, *The Classical Theory of Fields* (Pergamon Press, Addison-Wesley Publishing Company Inc., 1962), pp. 341–344.

9. Dyson [Freeman].

10. Satchel [John], physicist and philosopher of science.

11. The Richtmyer Memorial Lecture: Some Historical Notes, *American Journal of Physics* **37** (1969) 577. Conservation laws in general relativity and in the post-Newtonian approximations, *The Astrophysical Journal* **158** (1969) 45–54.

12. The second post-Newtonian equations of hydrodynamics in general relativity (with Y. Nutku), *The Astrophysical Journal* **158**, 55–79. The oscillations of a rotating gaseous mass in the post-Newtonian approximation in general relativity in *Quanta*, eds. P. G. O. Freund, C. J. Goebel and Y. Nambu (University of Chicago Press, 1970), pp. 188–95. Kip Thorne [Kip S. Thorne]; *The Astrophysical Journal* **158** (1969a) 997. Peres A., *Nuovo Cimento* **15** (1960) 351.

13. Trautman A., *Bull. Acad. Polon. Sci.* **6** (1958a) 627; also *Lectures on General Relativity* (1958b); mimeographed notes (London: King's College).

14. The $2\frac{1}{2}$ post-Newtonian equations of hydrodynamics and radiation reaction in general relativity (with F. P. Esposito), *The Astrophysical Journal* **160** (1970) 153–79. Post-Newtonian methods and conservation laws in *Relativity*, eds. M. Carmeli, S. I. Fickler and L. Witten (Plenum Press, New York), pp. 81–108.

V. The Fallow Period (1970–1974)

1. Carter [Brandon] an Australian theoretical physicist, noted for his work on black holes at the faculty of Meudon campus of the Laboratoire Univers et Théories. Geroch [Robert P.], noted Relativist, colleague at the University

of Chicago. Ellis [George F. R.], Cosmologist, co-author of *The Large Scale Structure of the Universe with Stephen Hawking.*

2. Friedman [John* (1973)]. Luyten [Willem J. (1899–1994)] Noted Astronomer, discoverer of many white dwarfs and the star named after him, the Luyten star. Persides [Sotirios C.* (1970)]. Norman [Norman Lebovitz].

3. J. R. Oppenheimer and H. Snyder, *Phys. Rev.* **56** (1939) 455.

4. The papers before leaving for India on March 31, 1970: The post-Newtonian effects of general relativity on the equilibrium of uniformly rotating bodies, V. The deformed figures of the Maclaurin spheroids (Continued), *The Astrophysical Journal* **167** (1971) 447–53. The post-Newtonian effects of general relativity on the equilibrium of uniformly rotating bodies, VI. The deformed figures of the Jacobi ellipsoids (Continued), *The Astrophysical Journal* **167** (1971) 455–63. Some elementary applications of the virial theorem to stellar dynamics (with D. Elbert), *Monthly Notices of the Royal Astronomical Society* **155** (1971) 435–47. A limiting case of relativistic equilibrium (in honor of J. L. Synge) in *General Relativity*, ed. L. O'Raifeartaigh (Clarendon Press, Oxford), pp. 185–99. Lady Raman [wife of Sir C. V. Raman (1888–1970)], Ramaseshan [Sivraj (1923–2003)], one of India's most accomplished scientists. Editor of C.V. Raman's papers on Light Scattering.

5. On the "derivation" of Einstein's field equations, *American Journal of Physics* **40** (1972) 224–34.

6. Hartle [James B.] American physicist, currently at the University of California, Santa Barbara, noted for his work on general relativity, astrophysics and interpretation of quantum mechanics.

7. Press [William], Teukolsky [Saul A.], both to become eminent scientists. Press after a distinguished career as a theoretical astrophysicist and Professor of Astronomy and Physics at Harvard University for twenty years, is currently at the University of Texas, Austin, as the chair in computer sciences and integrative biology. Teukolsky is a Professor of Physics and Astronomy at Cornell University. He is recognized for seminal contributions in general relativity, relativistic astrophysics and computational astrophysics.

8. On the stability of axisymmetric systems to axisymmetric perturbations in general relativity, I. The equations governing nonstationary, stationary, and perturbed systems (with J. L. Friedman), *The Astrophysical Journal* **175** (1972) 379–405. On the stability of axisymmetric systems to axisymmetric perturbations in general relativity, II. A criterion for the onset of instability in uniformly rotating configurations and the frequency of the fundamental mode in case of slow rotations, *The Astrophysical Journal* **176** (1972) 745–68.

9. Stability of stellar configurations in general relativity, Proceedings at Meeting of the Royal Astronomical Society, *The Observatory* **92** (1972) 160–74. The increasing role of general relativity in astronomy (Halley Lecture),

The Observatory **92** (1972) 160–74.

10. On the stability of axisymmetric systems to axisymmetric perturbations in general relativity, IV. Allowance for gravitational radiation in an odd-parity mode (with J. L. Friedman), *The Astrophysical Journal* 181 (1973) 481–95. Ehlers [Jürgen (1928–2008)], among the distinguished German physicists who specialized in General Relativity.

11. Wheeler [John A. (1911–2008)], Deser [Stanley]. P. A. M. Dirac on his seventieth birthday, *Contemporary Physics* **13** (1973) 389–94.

12. On a criterion for the occurrence of a Dedekind-like point of bifurcation along a sequence of axisymmetric systems, I. Relativistic theory of uniformly rotating configurations (with J.L. Friedman), *The Astrophysical Journal* **185** (1973) 1–18. On a criterion for the occurrence of a Dedekind-like bifurcation long a sequence of axisymmetric systems, II. Newtonian theory for differentially rotating configurations (with N. R. Lebovitz), *The Astrophysical Journal* **183** (1973) 19–30.

13. On a criterion for the onset of dynamical instability by a non-axisymmetric mode of oscillation along a sequence of differentially rotating configurations, *The Astrophysical Journal* **187**, 169–74. Sciama [Dennis W. (1926–1999)], British physicist, considered as one of the fathers of Modern Cosmology. On the slowly rotating homogeneous masses in general relativity (with J. C. Miller), *Monthly Notices of the Royal Society* **167** (1974) 63–79.

14. The black hole in astrophysics: The origin of the concept and its role, *Contemporary Physics* **14** (1974) 1–24. The deformed figures of the Dedekind ellipsoids in the post-Newtonian approximation to general relativity (with Donna Elbert), *The Astrophysical Journal* **192**, 731–46. Corrections and amplifications to this paper in *The Astrophysical Journal* **220** (1978) 303–11, are incorporated in the present version. T. Regge and J. A. Wheeler, *Phys. Rev.* **108** (1957) 1063.

15. F. J. Zerilli, *Phys. Rev. D* **2** (1970) 2141. J. M. Bardeen and W. H. Press, *J. Math. Phys.* **14** (1972) 7–19. S. A. Teukolsky, *Astrophys. J.* **185** (1973) 635. On the equations governing the perturbations of the Schwarzschild black hole, *Proceedings of the Royal Society A* **343** (1975) 289–98. The quasi-normal modes of the Schwarzschild black hole (with S. Detweiler), *Proceedings of the Royal Society A* **344** (1975) 441–52.

VI. General Relativity; Ryerson Lecture; Separation of Dirac Equation
(January 1975–August 1977)

1. On the equations governing the axisymmetric perturbations of the Kerr black hole (with S. Detweiler), *Proceedings of the Royal Society A* **345** (1975) 145–67; RYERSON LECTURE, Shakespeare, Newton, and Beethoven or patterns of creativity, delivered April 22, 1975. Worth noting the remarkable

effort in preparing for this lecture during the period of convalescence after a major heart surgery.

2. On a transformation of Teukolsky's equation and the electromagnetic perturbations of the Kerr black hole, *Proceedings of the Royal Society A* **348** (1976) 39–55. Remark about of Monique Tassoul finding an error. Verifying the Theory of Relativity, Notes and Records, *Roy. Soc.* **30** (1976) 249–260. Originally published in the *Bulletin of the Atomic Scientists* under the title "Of Some Famous Men."

3. Varena Lecture: "Why Are the Stars as They Are?"

4. On coupled second-harmonic oscillations of the congruent Darwin ellipsoids, *The Astrophysical Journal* **202** (1975) 809–14. The solutions of Maxwell's equations in Kerr geometry, *Proceedings of the Royal Society A* **349** (1975) 1–8. On the equations governing the gravitational perturbations of the Kerr black hole (with S. Detweiler), *Proceedings of the Royal Society, A* **350** (1976) 165–74. Second Varena Lecture: "On the linear perturbations of the Schwarzschild and the Kerr Metrics," (December 19, 1975).

5. Weil [André (1906–1998)], eminent mathematician renowned for his work in number theory and algebraic geometry. Hadamard [Jacques (1865–1963)], French mathematician.

6. Cartan [Élie (1869–1951)], noted French mathematician for his fundamental contributions pure as well as mathematical physics. Cartan had introduced the concept of "Torsion tensor" in 1932 and had generalized Einstein's theory of Relativity.

7. Remarkable story of the separation of variables, starting and finishing in the same evening, in Dirac's equations in Kerr geometry. The solution of Dirac equation in Kerr geometry, *Proceedings of the Royal Society A* **349** (1976) 571–75. Neutrino waves: On the reflection and transmission of neutrino waves by a Kerr black hole (with S. Detweiler), *Proceedings of the Royal Society A* **352** (1977) 325–38.

8. Pages 10–12 give a detailed account of trials and errors in achieving complete integration of the Newman–Penrose equations. Xanthopoulos [Basils C.*], who collaborated with Chandra almost continuously since 1978 till "... was shot to death in an unspeakable act of violence on the evening of November 27, 1990, while giving a seminar lecture at the Research Center of the University of Crete (Iraklion, Greece), ending without warning, "a life of love, joy, rich in Promise ... My association with Basilis is the most binding in all my sixty years in science." [S. Chandrasekhar, In remembrance of Basilis Xanthopoulos, Author's Note in Selected Papers, Volume 6].

9. Ernst's Equation; F. J. Ernst, *Phys. Rev.* **167** (1968) 1175–8; *ibid.* **168** (1968) 1415–7.

10. The Kerr metric and stationary axisymmetric gravitiational fields, *Proceedings of the Royal Society A* **358** (1978) 405–20; The gravitational pertur-

bations of the Kerr black hole, I. The perturbations in the quantities which vanish in the stationary state, *Proceedings of the Royal Society A* **358** (1978) 405–20; The gravitational perturbations of the Kerr black hole, II, *The Proceedings of the Royal Society A* **358** (1978) 441–65. The dates of submission of these papers were April 18, May 2, and June 20, all in 1977!

11. The deformed figures of the Dedekind ellipsoids in the post-Newtonian approximation to general relativity: Corrections and amplifications (with D. Elbert), *The Astrophysical Journal* **220** (1978) 303–13. An incident in the life of S. Ramanujan, F.R.S.: Conversations with G. H. Hardy, F.R.S. and J. E. Littlewood, F.R.S.; and their sequel (Archives of the Royal Society). On the discovery of the enclosed photograph of S. Ramanujan, F.R.S. (Archives of the Royal Society).

12. Edward Arthur Milne: recollections and reflections (Archives of the Royal Society). Book Review: *A History of Ancient Mathematical Astronomy* (3 volumes) by O. Neugebauer, *Bulletin of the American Mathematical Society* (with N. Swerdlow).

VII. General Relativity; Kerr–Newman Perturbations
(August 1977–December 1978)

1. As 1976 came to an end, Chandra felt his work on black holes was reaching a climax. During the spring quarter of 1977, while teaching a graduate course, he worked hard preparing three manuscripts [Ref. 10 in the previous chapter] and felt he was nearing a complete solution of the perturbation of the Kerr metric. But the work was proving to be exceptionally difficult and taking a heavy toll on his health with periodic experience of pressure in his chest. He learned that he had a serious heart problem, having had a heart attack in 1974. He was advised to have a test (cardio catheterization) to decide whether he needed a heart surgery. With the work near completion within a few months and also the General Relativity Conference (GR-8) just a few months away in August, he decided to postpone the test, lest it would lead to an immediate surgery, only after the return from the conference. In late August soon after his return, he underwent a major surgery with three bypass valves put in.

2. Israel [Werner] Noted Canadian physicist with seminal contributions to black hole physics. Pages 3–8 describe setbacks, problems and the role of several people in the completion of the following papers: The gravitational perturbations of the Kerr black hole, III. Further amplifications, *Proceedings of the Royal Society A* **365** (1979) 425–51. Moncrief, V, *Phys. Rev. D* **9**, 2707 and *D* **10** (1974) 1057. On the metric perturbations of the Reissner-Nordström black hole (with B.C. Xanthopoulos*), *Proceedings of the Royal Society A* **367** (1979) 1–14.

3. Einstein and general relativity: Historical perspectives (1978 Oppenheim Memorial Lecture), *American Journal of Physics* **47** (1979) 212–17. Einstein's general theory of relativity and cosmology, in *The Great Ideas of Today*, 90–138. Encyclopedia Britannica.

4. Matzner, R., *Phys. Rev. D* **14** (1976) 3724. On the equations governing the perturbations of the Reissner–Nordstöm black hole, *Proceedings of the Royal Society A* **365** (1979) 453–65.

VIII. A Year of Failures and Obligations (1979)

1. On the potential barriers surrounding the Schwarzschild black hole, in *Spacetime and Geometry: The Alfred Schild Lectures*, eds. R. A. Matzner and L. C. Shepley (University of Texas Press, Austin, 1982), pp. 120–46.

2. Beauty and the quest for beauty in science, *Physics Today* **32** (1979) 25–30. R. Narasimhan, a colleague in the Mathematics Department.

3. Monodromic Group; **Monodromy** is the study of how objects from *mathematical analysis, algebraic topology* and *algebraic* and *differential geometry* behave as they 'run round' a *singularity*. Monodromy Group is a group of transformations acting on the data that encodes what does happen as we 'run round' a singularity. Sorkin [Rafael], noted relativist, associate in the physics department.

4. On one-dimensional potential barriers having equal reflexion and transmission coefficients, *Proceedings of the Royal Society A* **369** (1980) 425–33. I.A.U meeting in Montreal talk: The role of general relativity in astronomy: Retrospect and prospect, in *Highlights in Astronomy*, Vol. 5, ed. P. A. Wayman (D. Reidel, Dordrecht, Holland), pp. 45–61.

5. General theory of relativity: The first thirty years, *Contemporary Physics* **21** (1980) 429–49; Oxford Lecture: 'Edward Arthur Milne: His part in the development of modern astrophysics', Unesco Lecture: Black holes: the why and the wherefore.

6. Pages 8–9. Discussion of his dissatisfaction with the previous paper on the Gravitational Perturbations of the Kerr black hole and final resolution and publication: The gravitational perturbations of the Kerr black hole, IV. The completion of the solution, *Proceedings of the Royal Society A* **384** (1980) 301–15. Ready to write the book *The Mathematical Theory of Black Holes*.

IX. The Mathematical Theory of Black Holes (1980–1981)

1. Wilkinson [Denys], Noted British Nuclear Physicist. Parker [Eugene], Distinguished Service Professor in the Department of Physics, Astronomy and Astrophysics.

2. References for 'Cotton-Darbough' theorem, so designated by Chandra, E. Cotton, *Ann. Fac. Sc. Toulouse, Ser.* **2**, 1 (1899) 385–438; G. Darboux,

Lecons sur les Systèmes Orthogonaux et les d Cordinnèes Curvilignes (Gathier-Villars, Paris, 1898).

3. Xanthopoulos theory: B. C. Xanthopoulos, *Proc. Roy. Soc. (London) A* **378** (1981) 61–71.

4. M. Walker and R. Penrose, *Commun. Math. Phys.* **18** (1970) 265–74; A specific type of space-time, type-D in Petrov classification, A. Z. Petrov, *Einstein Spaces*, translated by R. E. Kelleher and J. Woodrow (Pergamon Press, Oxford, 1969); For Kerr-Schild co-ordinates, see R. P. Kerr and A. Schild, *Proceedings of Symposia in Applied Mathematics* **17**, American Math. Soc. (1965) 199–209; B. Carter, *Phys. Rev. Lett.* **26** (1972) 331–33; D. C. Robinson, *ibid.* **34** (1975) 905–6.

5. P. A. Connors and R. F. Stark, *Nature* **269** (1977) 128–9. Penrose process of extracting the rotational energy of a Kerr black hole: R. Penrose and R. M. Floyd, *Nature Phys. Sci.* **229** (1971) 177–9.

6. Superradiance refers to a class of radiation effects (or enhanced radiation effects) typically associated with the acceleration or motion of a nearby body (which supplies the energy and momentum for the effect). Superradiance allows a body with concentration of angular or linear momentum to move towards a lower energy state, even when there is no obvious classical mechanism for this to happen.

7. Klein paradox refers to a surprising result found by Oscar Klein obtained by applying the Dirac equation to the familiar problem of electron scattering from a potential barrier. In non-relativistic quantum mechanics, electron tunneling into a barrier is observed, with exponential damping. However, Klein's result showed that if the potential is on the order of the electron mass, $eV \sim mc^2$, the barrier is nearly transparent. Moreover, as the potential approaches infinity, the reflection diminishes and the electron is always transmitted.

8. R. Geroch and J. B. Hartle, *J. Math. Phys.* **23** (1982) 680–92; J. B. Hartle and S. W. Hawking, *Commun. Math. Phys.* **26** (1972) 87–101.

9. Wald's procedure; R. M. Wald, *Astrophys. J.* **191** (1974) 231–3; *Ann. Phys.* **82** (1974) 548–56.

10. J. L. Friedman and B. F. Schutz Jr., *Phys. Rev. Lett.* **32** (1973) 243–5.

X. POSTSCRIPT: 1982 a Year that Passed

1. On Crossing the Cauchy Horizon of a Reissner–Nordström Black Hole (with J. B. Hartle), *Proceedings of the Royal Society A* **384** (1982) 301–15; Vikram Sarabhai Lectures: *Why Are the Stars as They Are? Mathematical Theory of Black Holes I & II*.

2. Eddington: *The Most Distinguished Astrophysicist of His Time* (Cambridge University Press, Cambridge, 1983).

XI. The Beginning of the End (1983–1985)

[The ensuing pages record Chandra's own list of publications, lectures and trips away from home. I refer to them in my Notes.]

1. Algebraically special Perturbations (1). Hayakawa [Satio (1923–1992)] Doyen of Japanese physics, leader in several branches of physics including cosmic rays, particle and nuclear physics and cosmology. Padua GR-10 Lecture (3)
2. Announcement from Stockholm. Paper with Norman Lebovitz (2). Khan, K. and Penrose, R., *Nature, Lond.* **229** (1971) 185.
3. Valeria Ferrari, Ruffin's student. Rufini [Remo], Professor of Theoretical Physics at the University of Rome "Sapienza" and the President of the International Center of Relativistic Astrophysics.
4. Nutku, Y. and Halil, M., *Phys. Rev. Lett.* **39** (1977) 1379.
5. Going to Rome to work with and complete the paper on the Nutku–Halil solution for colliding impulsive gravitational waves (3).
6. Paper with Xanthopoulos on colliding waves in the Einstein–Maxwell theory (4).
7. Cronin [James] Experimental nuclear and particle physicist, co-discoverer of CP violation in weak interactions of elementary particles. The pursuit of science and its motivations. Lectures (5).
8. Going to Crete to work with Xanthopoulos to get the work done within a week to complete the paper on the collision of impulsive gravitational waves when coupled with fluid motions (5, 6) and then rushing to consult with Roger Penrose. Penrose's suggestion leading to paper (6) on the collision of impulsive gravitational waves coupled with null dust with Xanthopoulos.

XII. Continued Effort I (September 1985–May 1987)

[As in the previous Chapter, numbers refer to Chandra's own record of publications, Lectures and trips.]

1. Visit to Crete to work with Xanthopoulos combined with brief vacation. Problem with paper (1): A new type of Singularity. Ablowitz [Mark J.] Professor in Applied Mathematics at the University of Colorado; coauthor of books on solitons. Rush to Oxford to consult with Roger Penrose.
2. Papers (1) and (2) completed, Solutions of the Einstein–Maxwell equations and generalizations of a solution by Bell and Szekeres: [Bell, P. & Szekeres, P., *Gen. Rel. Grav.* **5** (1974) 275.] On colliding waves that develop time-like singularities (3).
3. Schwarzschild Lecture. The aesthetic base of the general theory of relativity. Lectures (1).

4. Woltjer Lo [Lodewiljk] Astronomer, well-known for his studies on the Crab Nebula. Oort [Jan H. (1900–1992)] Noted Dutch astronomer, a pioneer in the field of radio astronomy.

5. Paper with V. Ferrari. On the dispersion of cylindrical impulsive gravitational waves (4).

6. Kothari [Daulat S. (1905–1993)], an eminent scientist from India. Balakrishnan and Shyamala [younger brother and sister-in-law]. Savitri [Mrs. R. R. Sarma] and Vidya [Mrs. V. V. Shankar] sisters.

7. Paper with B. Xanthopoulos. The effect of sources on horizons (5). Begin final work on Newton's Principia.

XIII. Continued Efforts II (May 1987–September 1989)

[In this chapter, Chandra has provided references to some of his papers in the text. I provide here only the missing ones and bibliographical references mentioned in the text]

1. Weyl, H., *Ann. Phys.* **54** (1917) 117; Neumann, F. E., *Beiträge zur Theorie der Kugelfunctionen* (Leipzig, 1878).

2. Flying to consult with Roger Penrose and completing the paper, A perturbation analysis of the Bell–Szeckeres space-time (with B. C. Xanthopoulos) *Proceedings of the Royal Society A* **420** (1988) 93–123. Majumdar–Papapetrou solution: Solution discovered independently by S. D. Majumdar, *Phys. Rev.* **72** (1947) 390–8; A. Papaptrou, *Proceedings of the Roy. Irish Acad.* **51** (1947) 191–205; Gibbons [Gary], theoretical physicist, professor at Cambridge University.

3. Winston [Roland] known for his path breaking research on Non-Imaging Optics in the field of Solar Energy. Rosner [Jonathan l.], Nambu [Yoichiro], noted theoretical particle physicists at the University of Chicago. The two-center problem in general relativity: The scattering of radiation by two extreme Reissner–Norström black holes, *Proceedings of the Royal Society A* **421** (1989) 227–58.

4. A one-to-one correspondence between the static Einstein–Maxwell and stationary Einstein–Vacuum space-times, *Proceedings of the Royal Society A* **423** (1989) 379–86; Two black holes attached to strings, with B. C. Xanthopoulos, *Proceedings of the Royal Society A* **423** (1989) 387–400.

XIV. Continued Efforts III (September 1989–October 1991)

1. Short paper: The account submission and withdrawal and resubmission form quite a fascinating account of never ending efforts at a solution of the problem: The flux integral for axisymmetric perturbations of static space-times (with V. Ferrari), *Proceedings of the Royal Society A* **428** (1990) 325–49; The

Einstein pseudo-tensor and the flux integral for perturbed static-space-times (with V. Ferrari), *Proceedings of the Royal Society A* **435** (1991) 645–57.

2. No less fascinating and dramatic is the effort at writing the next two papers:

3. On the non-radial oscillations of a star (with V. Ferrari), *Proceedings of the Royal Society A* **432** (1991) 247–79; On the non-radial oscillations of slowly rotating stars (with V. Ferrari), *Proceedings of the Royal Society A* **433** (1991) 423–40.

4. A long two-hour discussion with Roger Penrose on the topic of motivations in the pursuit of science and the aesthetic component. Pertinent articles: The Series Paintings of Claude Monet and the Landscape of General Relativity. Dedication Address, Inter-University Center for Astronomy and Astrophysics, 28 December 1992. Science and Scientific Attitudes, *Nature* **279**, no. 6264 (22 March 1990) 285–86.

5. On the non-radial oscillations of a star. II (with V. Ferrari and R. Winston), *Proceedings of the Royal Society A* **434** (1991) 635–41.

6. On the non-radial oscillations of a star. III. A reconsideration of the axial modes (with V. Ferrari), *Proceedings of the Royal Society A* **434** (1991) 449–57.

7. Regge Theory of Potential Scattering; Alfaro, V. de & Regge, T., Potential scattering (Amsterdam North Holland Press, 1963); On the non-radial oscillations of a star. IV. An application of the theory of Regge Poles (with V. Ferrari), *Proceedings of the Royal Society A* **437** (1992) 133–49.

XV. Continued Efforts IV (November 1991–December 1994)

[Attaching Chandra's own record of his activities with some details added]

1. On the occasion of the Charles Greeley Abbot Award by the American Solar Energy Society, *Solar Energy* **51**(3) (1993) 233–35; Chalonge Symposium: Daniel Chalonge and the Problem of the Abundance of Hydrogen. Opening Address, First Course of the International School of Astrophysics. "D. Chalonge," *First Course Current Topics in Astrofundamental Physics*, eds. N. Sánchez and A. Zichichi (World Scientific Publishing Co., River Edge, N.J., 1992) [*Selected Papers of S. Chandrasekhar*, Volume 7 (University of Chicago Press, 1997), p. 249].

2. Rayleigh–Mie Theory: Mie G., *Ann. Phys.* **4** (1908) 377; Rayleigh, Lord, *Phil. Mag.* **61** (1871) 447–454. (Also Scientific Papers (reprinted in Dover Publications (1964), vol. 1, p. 104). On spherical free gravitational waves (with V. Ferrari), *Proceedings of the Royal Society A* **443** (1993) 445–49.

3. The Series Paintings of Claude Monet and the Landscape of General Relativity. Dedication Address, Inter-University Center for Astronomy and Astrophysics, **28**, December 1992. [*Selected Papers of S. Chandrasekhar*, Volume 7, p. 127].

4. Newton and Michelangelo, *Current Science* **67**, no. 7 (10 October 1994) 497–99. [*Selected Papers*, Volume 7, p. 241]. On reading Newton's *Principia* at age past eighty, *Current Science* **67**, no. 7 (10 October 1994) 495–96 [*Selected Papers*, Volume 7, p. 235].
5. On the non-radial oscillations of a star: V. A fully relativistic treatment of a Newtonian star (with V. Ferrari), *Proceedings of the Royal Society A* **450** (1995) 450–76. "The Equations were so beautiful that I was convinced that the approach must be right."